Fighting for Farming Justice

This book provides a detailed discussion of four class-action discrimination cases that have recently been settled within the United States Department of Agriculture (USDA) and have led to a change in the way in which the USDA supports farmers from diverse backgrounds. These settlements shed light on why access to successful farming has been so often limited to white men and/or families, and significantly this has led to a change for opportunities in the way the USDA supports famers from diverse backgrounds. With chapters focusing on each settlement, Jett provides an overview of the USDA before diving into a closer discussion of the four key settlements, involving African American farmers (Pigford), Native Americans (Keepseagle), Woman famers (Love) and Latino(a) farmers (Garcia), and the similarities between each. This title places an emphasis on what is happening in the farming culture today, drawing connections between these four settlements and the increasing attention on urban farming, community gardens, farmers markets, organic farming and the slow food movement, through to the larger issues of food justice and access to food.

Fighting for Farming Justice will be of interest to scholars of food justice and the farming arena, as well as those in the fields of Agricultural Economics, Civil Rights Law and Ethnic Studies.

Terri R. Jett is a Professor of Political Science, Butler University, Indianapolis, USA. She is also Special Assistant to the Provost for Diversity and Inclusivity an affiliated faculty with the Peace and Conflict Studies Program and the Gender, Women, Sexuality Studies Program.

Other books in the Earthscan Food and Agriculture Series

Multifunctional Land Uses in Africa
Sustainable Food Security Solutions
Elisabeth Simelton and Madelene Ostwald

Food Security Policy, Evaluation and Impact Assessment
Edited by Sheryl L. Hendriks

Transforming Agriculture in Southern Africa
Constraints, Technologies, Policies and Processes
Edited by Richard A. Sikora, Eugene R. Terry, Paul L. G. Vlek and Joyce Chitja

Home Gardens for Improved Food Security
Edited by D. Hashini Gelhena Dissanayake and Karimbhai M. Maredia

The Good Farmer
Culture and Identity in Food and Agriculture
Rob J.F. Burton, Jérémie Forney, Paul Stock and Lee-Ann Sutherland

Deep Agroecology and the Homeric Epics
Global Cultural Reforms for a Natural-Systems Agriculture
John W. Head

Fighting for Farming Justice
Diversity, Food Access and the USDA
Terri R. Jett

For more information about this series, please visit: http://www.routledge.com/books/series/ECEFA/

Fighting for Farming Justice
Diversity, Food Access and the USDA

Terri R. Jett

LONDON AND NEW YORK

First published 2021
by Routledge
2 Park Square, Milton Park, Abingdon, Oxon OX14 4RN

and by Routledge
52 Vanderbilt Avenue, New York, NY 10017

Routledge is an imprint of the Taylor & Francis Group, an informa business

British Library Cataloguing-in-Publication Data
A catalogue record for this book is available from the British Library

Library of Congress Cataloging-in-Publication Data
A catalog record has been requested for this book

ISBN: 978-0-367-00160-5 (hbk)
ISBN: 978-0-429-40105-3 (ebk)

Typeset in Times New Roman
by KnowledgeWorks Global Ltd.

I dedicate this book to the most brilliant farmer I know, my grandfather Rafe Taylor, Sr., as well as my grandmother Elnora Taylor, equally as brilliant. I also dedicate this book to my paternal grandparents, Earl and Ella Jett who inspired me with their creative genius, my parents Kenneth and Beatrice Jett, both life-long learners and hard workers who never cease to amaze me and my children, Akilah and Talib Shahid who inspire me to do better. I come from a beautiful extended family network, Taylor, Jett, Demery, Jones which has continued to ground me in a love of my people and all of the struggles that we have historically and continue to face with incredible resilience. The cultivation of food and life has been a hallmark of the challenges we have overcome and it is at the moment of writing this book that I hope that a recognition of the power of the marginalized and oppressed will finally be acknowledged and uplifted. All farmers are truly the backbone of every society.

Contents

List of Photos

List of Figure

List of Tables

Acknowledgements

I would like to acknowledge the tremendous support I have received in both my personal and professional life that allowed me the time and space needed to write this book that I hope will inspire many to recognize that land is power but that it must be shared, cultivated, conserved and sustained to benefit all for generations to come. I am thankful for all the support of my institution, Butler University that provided me with a fellowship during my sabbatical and for the support from a student apprentice, Naomi Norris who provided me with a tremendous amount of research on the Census of Agriculture and shared my enthusiasm for the data and topic. I have wonderful colleagues of whom I cannot name them all but my writing partner, Dr. Terri Carney was key in helping me focus and in her encouragement and friendship throughout my entire academic career. I also recognize that I am fortunate to "live" in the Department of Political Science/Peace and Conflict Studies at Butler University, where my interdisciplinary perspective has never had to fit in a traditional box.

My life partner, Daryl Campbell who patiently indulges my reading out loud of my work and is always a good listener but an even better traveling companion. My daughter, Akilah Shahid, one of the smartest people I know, provided me with tremendous support with her expert charting and graph skills.

Ultimately, the life of the Black intellectual must be one of the pursuits of liberation.

1 Introduction

Farming and agriculture have always been central to the development of this country and the lives of many families in the United States regardless of background even when many of these families migrated to various regions of the country to escape oppressive conditions and/or to seek greater economic opportunity. Today, still as much as 40% of the U.S. is farmland and about 35 million people are connected to the field of agriculture, whether farming their own land, renting land to farm, or working as farmworkers (*About the U.S. Department of Agriculture | USDA*, n.d.). At the heart of this foundation is the ongoing issue of sustainability in relationship to food and the ability of collective groups of people to be self-sufficient in the context of a global capitalist system. While the focus of this book is on the United States agricultural arena, we must additionally consider how it connects to a global food system. Eric Holt-Gimenez points out,

> The particular role of agriculture in capitalist development was addressed by classical political economists in seminal publications like "The Wealth of Nations," "An Essay on the Principles of Population" "The Principles of Political Economy and Taxation," and "Das Kapital." Economists like Adam Smith and David Ricardo concentrated on the nature of wealth creation, the market, and the differences of power between workers, peasants, landowners, and industrialists. Their concepts of property and commodities, the labor theory of value, land rent, and the creation of surplus value are still foundational to understanding the capitalist agriculture."

> (Holt-Gimenez, 2017)

Land ownership from the very beginning of this republic dictated access into decision-making circles that determined the structure of the United States indicating a certain level of status and, therefore, political and economic power. It was also seen as a means of both sustenance and entrepreneurship with agriculture the primary determinant of progress up until the 20th century and still holds much significance today. It was agriculture that was at the heart of our path toward the industrial revolution, and while we are currently in the midst of the technological revolution, agriculture is still valued and important. In fact, the use of technology now serves as one measure of access and success in farming (*USDA - National Agricultural Statistics Service - Census of Agriculture,* n.d.). Farmers, especially young farmers use social media for marketing and making connections to other related efforts such as farmer's markets and Community Supported Agriculture, and there are a number of individuals who belong to farming advocacy groups that exchange information amongst one another via social media.

The geographical expansion of the United States, especially the western land expansion in the 19th century with policies of aggressive homesteading measures set the stage for the 1862 establishment of the United States Department of Agriculture (U.S.D.A.), the "People's Department," as it was called right from the beginning by President Abraham Lincoln. However, the U.S.D.A. wasn't necessarily of service to all people at its beginnings and neither was the United States' efforts of land expansion. It is no surprise that the U.S.D.A. was established during the time period that the country was in the midst of wrestling with both the economical and moral implications of the slavery system.

> "Prior to slavery, capitalist agriculture failed to keep up with the growing demand for cotton because capitalists couldn't force the peasantry to grow it on an industrial scale. In the southern United States, settlers had exterminated and driven off indigenous populations to appropriate their land, a strategy that left them without a workforce. The enslavement and translocation of Africans from West Africa to North America and the Caribbean was capitalism's answer to the labor shortage."
>
> (Holt-Gimenez, 2017)

The initial establishment of the United States was built on the acquisition of federal land. Federal control of the land meant that the government dictated its usage, distribution, and conservation. Political and economic power had to be wedded from a national position in or for progress to continuous.

> Federal land ownership began when the original 13 states ceded their "western" lands (between the Appalachian Mountains and the Mississippi River) to the central government between 1781 and 1802. Substantial land acquisition in North America via treaties and purchase began with the Louisiana Purchase in 1803 and culminated with the purchase of Alaska in 1867. In total, the federal government acquired 1.8 billion acres in North America.
>
> (Alexander, 2007)

In the U.S.D.A. Economic Research Service Report, "U.S. Farmland Ownership, Tenure and Transfer," Bigelow, Borchers, and Hubbs (2016) write about the significance of farmland accessibility that has continued to shift to a mixture of rented and owned land that "have important implications for land accessibility, particularly for young and beginning farmers." How land has been maintained, transferred, or lost between generations is part of both the historical narrative and the contemporary struggles that define the agricultural arena. Land determined access to power so much so to the extent that many groups were prevented from accumulating land and/or deliberately had their land taken away through a number of government policies and/or through violent acts in efforts to maintain a predominantly white status quo and to keep more racial- and ethnic-identified marginalized populations in conditions of relative servitude. Historically, this often occurred through racial and class lines in an intersectional manner that largely explains some of the economic inequities that exist today. And it is because of this reason, the agricultural arena and farming, in particular, have a significant social justice component that is at the forefront today, tied to the struggle of a number of these marginalized groups that have especially had disparate relationships with the U.S.D.A. Their ability to farm was key to their survival and the intentional disruption of their efforts has been devastating for many communities.

This is the primary focus of this book – the connection between their struggles to farm in relation to the food justice movement today. How have these groups managed to navigate the systemic policies and structural impediments that were at the very foundation of our agricultural development and what are the possibilities, though they may be limited for transformative and radical changes to take place? It is through understanding the respective cases of the Black, Native American, Latino/Hispanic and women farmers against the U.S.D.A. that help provide critical insight for what needs to occur going forward.

> The U.S. Department of Agriculture (USDA) has long been accused of unlawfully discriminating against minority and female farmers in the management of its various programs, particularly in its Farm Service Agency loan programs. Meanwhile, some minority and female farmers who have alleged discrimination by USDA have filed various lawsuits under the Equal Credit Opportunity Act (ECOA) and the Administrative Procedure Act (APA). Pigford v. Glickman, filed on behalf of African-American farmers, is probably the most widely known, although Native American and female farmers also filed suit in Keepseagle v. Vilsack and Love v. Vilsack, respectively. In addition, a group of Hispanic farmers filed a similar lawsuit against USDA in October 2000. The case, Garcia v. Vilsack, involved allegations that USDA unlawfully discriminated against all similarly situated Hispanic farmers with respect to credit transactions and disaster benefits in violation of the ECOA, which prohibits, among other things, race, color, and national origin discrimination against credit applicants. The suit further claimed that the USDA violated the ECOA and the APA by systematically failing to investigate complaints of discrimination, as required by USDA regulations.
>
> (Feder and Cowan, 2013)

My grandfather, Rafe Taylor, Sr., had all the markings of a displaced farmer as a young man of about 30-years-old who migrated from Bossier City, Louisiana to Oakland, California. At the time of his migration, he had a wife, Elnora and five children – (Mildred, Bea, Rafe, Jr., Laura and Amos – though Laura would tell people she was born in "San Francisco"), which he brought to Oakland a year after he was settled, first securing an apartment and then by the time his family came, he and his wife purchased their first home on Filbert Street. Because his father, William Taylor, was so dependent on his labor for their Louisiana farm, my grandfather Rafe Taylor, Sr., gave his father a mule to replace him when he left. After all, he said he "worked him like a mule."

The time period of his migration was during World War II and his initial job was in the Iron Foundry in Oakland, and after a few years, he was able to secure a position with the City of Oakland as a sewer mechanic, a job he worked at for over 30 years until he retired.

Photo 1.1 Rafe & Elnora Taylor at Meldon Avenue Home.

Photo 1.2 Rafe and Elnora Taylor – Wedding Day.

Photo 1.3 Rafe and brother, Flenore Taylor in front of school in Baton Rouge, Louisiana.

Photo 1.4 Rafe and Elnora Taylor well established.

Photo 1.5 Elnora Taylor on first day of work at Clorox, Inc.

Photo 1.6 Rafe and Elnora Family portrait with all of the children.

Though he was highly skilled, the level of his educational attainment limited his ability to move up with the department. His wife, Elnora migrated a year after Rafe, Sr., had three more children and also worked initially for the Cannery, Granny Goose Potato Chips, and then Clorox, Inc., until she retired, after about 20 years.

A few of Rafe Taylor, Sr.'s 14 brothers followed his lead in migrating to Northern California, a few of them after serving in the military during World War II, with the others and his sisters following later and after their father, William Taylor died. Their mother, Minerva Taylor, was the last to come. "Why did you all leave Louisiana?" I remember asking him one day, shortly before he passed away. He responded, "Because we were tired of having to cross the street every time there was a white person walking toward us on the same side of the street." He also was looking for better economic opportunities and wanted his children to have a better education then he had in Louisiana.

During one of our family reunions, which took place in Natchitoches, Louisiana my grandfather spotted a decrepit log cabin sitting on top of a grassy hill. He took off running up the hill as if he hadn't just been walking down the road with a cane, many of his grandchildren, including myself running after him. What moved him to react so passionately was that he said the log cabin was the actual location where he and his brothers had gone to school – for him, up to what would be the equivalent of the sixth grade. "We got a good education there," he stated. In reference to his comment about why he left Louisiana, this was his way of telling me that racism was too much for them to handle daily. Regardless, they were a farming family in Louisiana, a family of 17 children total, and so he and all of his siblings naturally carried their farming skills with them to the very urban environment of Oakland, California.

When Rafe and Elnora Taylor settled into their first house on Filbert Street in West Oakland, he dug up the large front yard lawn and planted a garden right in his front yard with greens, tomatoes and beans among other things, according to my mother, Beatrice Jett and my aunt Laura Johnson, his daughters. His garden was naturally abundant and overflowing and because it was in the front yard, at night people in the neighborhood would come by and help themselves, not realizing or perhaps disregarding that this was a matter of sustenance for his wife and children. Fortunately, the people who lived next door to them, one of the remaining few white families in that area, held a significant portion of the backyard land that included a Blackberry patch that the Taylor children would enjoy. They valued and appreciated Rafe Taylor, Sr.'s gardening/farming talents and also wanted to help so they allowed him to use their backyard land for his garden where he grew collard greens, tomatoes, and an abundance of other vegetables.

After some years, they purchased a second home in the Maxwell Park area of Oakland with a large backyard full of fruit trees and an expansive area to include a vegetable garden. In this backyard, Rafe Taylor, Sr., transformed a relatively small area by farming standards where he planted everything according to the placement of the sun and managed to feed his large family of eight children and countless grand- and great-grandchildren with what he grew with his own hands, organically. Fruit trees were at the top, apple, lemon, and fig, next was a layer of both collard and mustard greens. Then the tomatoes followed as another lower layer beneath the greens and at the very bottom were artichokes, which seemed to thrive in the direct sunlight. There was always a container on the kitchen counter where they collected composting material from kitchen scraps before it became a "process." It was regenerative, though I never heard my grandfather use that particular term – from the garden to the table – and back to the garden. His extensive garden may have also been considered "organic." The pathway to his house in the smallish front yard was lined with rose bushes – which he named after the important women in his life – Minerva (his mother), Lucy (his wife's mother) and Elnora (his wife). I often wondered how different things would be if our family had been comfortable remaining on the red soil that was so life-affirming in Louisiana.

So many Black families have similar stories of displacement and migration; a time period known as "The Great Migration," with patterns of movement that began during the Reconstruction period but reached a significant peak around the 1930s to 1940s driven largely because of the racism that existed at the time, decisions made to uproot, hoping for new economic opportunities and the freedom to succeed. Like my grandfather, they also brought their farming skills and means of survival from rural to urban environments. They carried the trauma of racism and injustice that had kept them in oppressive conditions of which they worked hard to survive, always thinking of better possibilities for future generations. And yet, there were also many Black families and especially Black farmers who remained in the deep south, that were critical to the success of the Civil Rights Movement while organizing and strategizing to hold on to their land against all types of adversity. People often generally think that most

Black people in the United States live in urban areas, and many do but there are many who live in rural Black Belt areas of the southern region. The term "Black Belt," is used to define the richness of the soil but also the predominance of Black populations at the county level. In Alabama, for example, there is a group of counties in the Southeastern part of the state that makes up what we could call "the Black Belt."

And thus, this book is driven by a premise – that out of the many struggles of farmers of diverse backgrounds and especially those who formed the basis for four alleged discrimination lawsuits and subsequent settlements against the U.S.D.A., known as Pigford I and II, Keepseagle, Garcia, and Love, came contemporary and imaginative ways of addressing food apartheid, food insecurity and food deserts in a movement of food justice efforts and forging more creative aspects of an ever-transformative agricultural and farming arena.

What is meant by terms such as food insecurity, food sovereignty, food apartheid and food justice? It is important to be clear on the specifics of these terms to draw connections between the struggles of the farming groups mentioned and the agricultural/farming arena today. Some aspects of these struggles also connect to issues surrounding climate change, environmental justice or racial ecologies that consider the ability or constraints surrounding the use of biodiversity practices of diverse communities that allow for organic farming practices to facilitate growth. While this text is focused specifically on the area of farming and food justice, no social justice issue can be looked at in a compartmentalized manner because there are so many factors that can be used to help explain disparities. Therefore, while the specific terms of food justice, food sovereignty, food apartheid and food insecurity will be defined, other facto will be used to explain disparate occurrences.

In their book, Food Justice, (2010) Robert Gottlieb and Anupama Joshi state that for the purpose of their book

> "in which food justice represents the substance as well as the governing metaphor of the discussion, we identify food justice in two ways. First, and most simply, we characterize food justice as ensuring that the benefits and risks of where, what, and how food is grown and produced, transported and distributed, and accessed and eaten are shared fairly. Second, by elaborating what food justice means and how it is realized in various settings, we hope to identify a language and a set of meanings, told through stories as well as analysis, that illuminate how food injustices are experienced and how they can be challenged and overcome."
>
> (Gottlie and Joshi, 2010)

In this case, the ability to farm successfully must be addressed in the context of both national and global food systems that are influenced by so many externalities that are beyond the control of the individual or farming collective and require some type of governmental intervention unless it is the government(s) that are, in fact, causing injustices. In which case, the ability to affect change in a food system to allow for greater participation requires activists and non-governmental organizations to insist on changes. According to Saryta Rodriguez in her edited book Food Justice: A Primer, *Food Justice* is defined as "the belief that food is a basic right of all people." In the Universal Declaration of Human Rights, Article 25 (1) states, "Everyone has the right to a standard of living adequate for the health and well-being of himself and of his family, including food, clothing, housing and medical care and necessary social services, and the right to security in the event of unemployment, sickness, disability, widowhood, old age or other lack of livelihood in circumstance beyond his control." (*Universal Declaration of Human Rights*, 1948)

Food Sovereignty is defined as "A population's right to determine how it is fed," and *Food Insecurity* is defined as "A household-level economic and social condition of limited or uncertain access to adequate food." Rodriguez further explains that when looking at the "Food Justice Movement," both food insecurity and food sovereignty are of primary concerns. For the purposes of this text, all contemporary activities related to farming today that have at its core a mission to be accessible and inclusive and to address any past or present disparities related to food and sustainability will be seen as being a part of the food justice movement (Rodriguez, 2018). Tribal food sovereignty is specifically defined as the ability of an indigenous nation to determine and control its food sources and development specifically without other governmental intrusions (*Tribal Food Sovereignty Definition*, n.d.). Dorceta E. Taylor states,

> Food justice and food sovereignty are narrative frames that occupy critical spaces in the discourses about food production and sustainability. Food justice and food sovereignty discourses combine interest in sustainability and consumption of healthy food with concerns about social justice, equitable access to healthy foods and control over the production of said food.
> (Taylor, 2018)

Tribal food sovereignty is particular to the Keepseagle case and the experiences of Indigenous nations in the United States and because it is so tied to U.S. public policy and treaty relations with the government. According to the International Indian Treaty Council, "Food sovereignty is the right of Peoples to define their own policies and strategies for sustainable production, distribution and consumption of food, with respect for their own cultures and their own systems of managing natural resources and rural areas, and is considered to be a precondition for Food Security" (*Food Sovereignty and the Rights of Indigenous Peoples*, 2013). And according to the "Declaration of Atitlan," from the 1st Indigenous People's Global Consultation on the Right to Food and Food Sovereignty held in Guatemala in 2002, "The rights to land, water, and territory, as well as the right to self-determination, are essential for the full realization of our Food Security and Food Sovereignty" (*Tribal Food Sovereignty Definition*, n.d.).

Food apartheid is a term defined by a number of food justice organizations to define the strategies they are using to address the deliberate racialized and economic policies that affect communities of color (Nishime & Williams (eds), 2018). There are a number of local decisions that affect communities and neighborhoods of color that disrupt their ability to sustain quality food production and access options. For example, decisions regarding community development that result in the displacement of people historically connected to redlining policies but in contemporary terms also known as gentrification. Or decisions to allow toxin-producing factories to set up in close proximity to lower-socioeconomic and/or communities of color that result in poor water and soil quality that prevent them from growing safe and healthy food.

The U.S.D.A. defines a food desert as a "low access community" where there are at least 500 people and/or 33% of the census tract's population must reside more than one mile from a supermarket or large grocery stores and for rural areas the distance is 10 miles (*USDA Defines Food Deserts | American Nutrition Association*, n.d.). The U.S.D.A. also provides a Food Locator map and a Food Atlas that indicates specific information as to where these food deserts exist (*USDA ERS - Food Access Research Atlas*, n.d.). As evidenced by the effect of racialized disparities that have occurred in the United States food system, a disproportionate number of neighborhoods and communities that are food deserts have concentrated populations of people-of-color. This is also the location of and focus of many urban farm efforts that have been deliberately started in those specific areas as a means to address the lack of available healthy food choices and to empower those specific communities to collectively becoming self-sufficient.

Regarding the four cases that are at the heart of this text, each of these groups sought opportunities as farmers to be self-sustaining and to grow their own capacity to address any type of food insecurities that may have existed within particular communities of which they especially identify. Unfortunately, these groups ran into institutional and structural impediments that deliberately prevented them from being successful in addition to other challenges that often face farmers regardless of their background, such as environmental and climate conditions. It is their ability to recover from these occurrences, especially in the case of natural disasters where there have additionally been governmental practices toward these farmers who have also served to be problematic and particularly mentioned in the Garcia settlement. With the Black, Native American, and Latino farmers, specifically, there is a historical context that provides an explanation as to why they had to seek both legal and activist recourse to force the system to change. It is through the descriptions of the structure and operations of the U.S.D.A. and its specific agencies that deal with farming, that the societal impediments to these farming groups will become apparent. Yet it is possible to look forward and imagine food systemic changes that can and may have occurred at the U.S.D.A. where there is an alignment of their priorities and policies with the food justice movement. Some of this has occurred merely because of the breadth of services that the U.S.D.A. provides to address poverty generally while others as a result of the settlements.

Some would argue that these efforts have expanded the arena and made it both more accessible and sustainable when it comes to the larger questions of farming and food production and the more deliberate focus of access to healthy food. These struggles and lawsuits served as the foundation for many contemporary practices that a number of younger, innovative farmers and farm collectives have used to make a living while simultaneously addressing the many current challenges of food insecurity throughout the country. Even further, these efforts have become a part of a global focus on solutions toward eradicating food hunger.

Some of these challenges are a result of federal policies that have changed agriculture from its original intention of primarily supporting predominantly white family farm homesteads to what many would argue is the current emphasis as a larger multinational agribusiness supporting agency. Other challenges are due to natural disasters and economic instability that have affected farmers and even more are a result of how communities, rural, suburban and urban have changed in relation to food access and distribution. And yet, regardless of background, these newer farmers are forming collectives, bridging the urban-suburban-rural divide and approaching farming with more environmentally sound practices, such as various no-till farming and organic farming practices, with sensitivity to the sustainability of the soil, and water preservation and other soil quality strategies that are doing so from an inclusive standpoint with a sensitivity to the effects of toxins on the lives of farmworkers. Subsequently, they are receiving some key support from a more accessible U.S.D.A., part of which happened as a result of discrimination lawsuits, though there are still some remaining challenges navigating the agencies' policies and political underpinnings. The settlement process itself, especially with Pigford and Keepseagle was wrought with challenges.

In addition to detailing the aforementioned settlements, this book will provide some noteworthy examples of these newer farmers and others in the agricultural arena and food justice movement who are transforming the field of agriculture and opening up a paradigmatic shift that makes this arena more accessible and far-reaching. It will include the perspectives and innovations of those who are specifically a part of these farming groups and have suffered discrimination and those who have embraced new progressive forms of farming that are centered in a farming and food justice identity, regardless of race, gender, economic status or many other determinants, such as former incarceration. This is important because there is value in understanding the complexity of nation-building from the very practical, personal and political standpoint that farming provides – those who benefitted from governmental practices and those who suffered and why.

What is currently happening in farming is a resurgence across communities to reclaim the value of the "family farm," or even just a transformed small farm, which is the predominant type of farm where even the definition of what constitutes the family has evolved. Further, technology and more specifically the way we communicate and share information has seen a revolution in the last few decades, and the practice of farming has also changed as a result, especially when it comes to the work of younger farmers. For example, many young farmers use social media to market their produce, reaching out to people through their organizational connections who may have a special interest in buying from them – such as vegans and vegetarians. In fact, this is now something that the U.S.D.A. is tracking in their Census of Agriculture, the number of farmers who are technologically savvy, though the use of technology could also include producing manipulation and not just marketing. And then there is the shift in farming global engagement such as trade agreements, but also immigration patterns that have most recently had a significant effect on food distribution and access. In a report from the U.S.D.A. Economic Research Service, authors James M. MacDonald, Robert A. Hoppe and Doris Newton detail how there has been a consolidation, a shift to larger farms in the last three decades, even while there has been an increase in smaller family farms. For example, they state, "By 2015, 51 percent of the value of U.S. farm production is from farms with at least $1 million in sales compared to 31 percent in 1991 (adjusted for price changes)" (MacDonald, Hoppe, & Newton, n.d.). They also state that "Despite increased consolidation, most production continues to be carried out on family farms, which are owned and operated by people related to one another by blood or marriage. Family farms accounted for 90 percent of farms with at least $1 million in sales in 2015 and produced 83 percent of production from million-dollar farms (MacDonald et al., n.d.)

As stated previously, in the United States of America land and land acquisition has always been an important mechanism of power, and yet for many people, it has also served as a means to survival. President Lincoln established the U.S.D.A. during a tenuous historical time period, the Civil War, and ever since then it has taken on and shaped its policies in a way to meet shifting demands and government aspirations unlike any other governmental agencies. The experiences of marginalized groups with the U.S.D.A. has always fit squarely in the context of larger historical societal strife so it is not surprising that these experiences culminated in a series of contemporary lawsuits and settlements that serve as some level of guidance to how farming and food justice plays out and change occurs.

The four settlements of collective farmer groups that shape the context for the premise that I have suggested are all versus (Tom) Vilsack, the U.S.D.A. Secretary at the time these settlements concluded under the administration of President Barack Obama. However, these cases were also handled by U.S.D.A. Secretaries Ann Veneman, under President George W. Bush, and Dan Glickman, under Presidents Bush and Clinton. However, the discriminatory experiences of these groups, pre-dates each of those U.S.D.A. secretaries and we can look to the policies of the U.S.D.A. under other secretaries to understand what ultimately led to these groups seeking legal action. The group terms (e.g., Black or Hispanic) used as identifiers are chosen to deliberately associate with the terms used in the cases at the time of their adjudication. These identifying terms will also correspond with the manner that the U.S.D.A. tracks these farmer groups through the Census of Agriculture with an understanding that many farmers of these categories did not regularly participate in this census for fear of having their land taken away from them, a fear justified from their historical governmental experiences, which will be discussed.

The contemporary plight of Black farmers first came to my attention when I saw John Boyd, Jr., President of the National Black Farmers Association featured as the "ABC Person of the Week," with Peter Jennings on November 21, 2003. He had driven an old-style wagon pulled by his mule, named "48 acres" from his farm in Baskerville, Virginia to the front of the U.S.D.A. to draw attention to the plight of Black farmers and their specific struggle with that governmental agency. Along the way, he picked up an additional mule named "Justice." In this news feature, Boyd offers, "I don't understand for the life of me that the good people on capitol hill can put laws in place to protect the bald eagle, the rockfish and I don't get that kind of reception to keep in place the oldest occupation in history for Black people in this country... which is farming." He did manage to speak to a few representatives once there including Representative Maxine Waters (D-CA – 43rd District) and Representative Dennis Kucinich (D-OH-) who at that time was a candidate for the Democratic nomination for President of the United States. Once I looked him up further, I discovered the case of the Black farmers. In addition to his activism work with Black farmers, John Boyd, Jr., was also instrumental in helping with what is known as the Cobell Settlement, signed by President Barack Obama in 2010 for $3.4 billion, which included a $1.9 billion Trust Land Consolidation Fund and a payment of $1.5 billion and affected over 300,000 individual American Indian Trust accounts (Boyd, 1995).

At the time when I was watching John Boyd, Jr., give rise to this issue, I had some knowledge of the historical plight of Black farmers and Black people generally who suffered great land loss over time and faced many impediments to farming self-sufficiency but had no idea that there was a pending civil case: Pigford v. Glickman. This case involved the experiences of Black farmers between the years of 1981 and 1996 with the U.S.D.A. with the named plaintiff, Timothy Pigford along with 400 other Black farmers. I then became aware of other subsequent civil cases against the U.S.D.A. included Native American farmers represented by George and Marilyn Keepseagle from the years 1986 to 1999, women farmers represented by Rose Marie Love from 1981 to 1996 and Latino farmers represented by Guadalupe L. Garcia (Garcia and Sons) from 1981 to 1996.

Each of these groups Pigford (Black farmers), Keepseagle (Native American farmers), Love (women farmers) and Garcia (Latino farmers), suffered discriminatory experiences with the agencies of the U.S.D.A. operating at the local level who were responsible for providing farm loan or loan servicing and other support for farmers during the specified time period. These practices were a systemic part of the U.S.D.A. primarily at the local level with no federal level accountability, largely due to the manner in which those who controlled the access to farming support were chosen through local county elections, so it was easy for structural racist and sexist barriers to become an entrenched part of the system as they were already a part of the existing community dynamic. What became even more egregious is that the process for redress of these structurally racist conditions was eliminated with the changes of Presidential administrations and the elimination of a lot of key federal oversight departments, such as civil rights divisions that may have been able to hear complaints from these farmers. These specific cases were listed as v. Veneman, v. Glickman and finally v. Vilsack as indicated in the varying and subsequent administrations involved throughout this process.

So discriminatory practices that prevented minority and women farmers from succeeding became almost expected to the detriment of many of their farms. Therefore, the stagnation and absolute decline of people of color and women to cultivate the land for food has been systematically historically woven into their oppression. Any means to change that system was thwarted at every level, local, state and federal. This is not to suggest that there were no efforts by these groups to change or even work around these oppressive conditions because this book will look to making the connections between those efforts and current farming and food justice-oriented practices. As stated before, the land is a source of power and independence and, therefore, any attempt toward more wealth, sustainability and/or land distribution, and cultivation was seen as a threat to the very structure of the United States, which is steeped in white male patriarchy, power and dominance.

Their civil right challenges to what was experienced with the U.S.D.A. had similarities in how their settlements were ultimately decided and there were some unintended consequences that arose from the processes put in place to actually decide on how the settlements would be distributed. The movement of these civil rights cases through the adjudication process, which involved the United States Department of Justice (U.S.D.o.J,) spanned over at least two and sometimes three administrations, further complicating matters. However, in addition to these settlements, these cases led to some changes in the structures of the U.S.D.A. and opened up access to governmental farming support that was previously denied. This book will provide details regarding these cases in addition to the changes that have been made in the U.S.D.A. and further draw connections to the current localized focus in many communities on food access and how to address food deserts in the context of a farming and food justice movement, some of which may have ironically been created by these discriminatory practices against the aforementioned groups in the first place.

With so much recent focus on the challenges of food insecurity in many communities across the United States, one can only imagine that had each of these groups not experienced deliberate structural racist and sexist impediments to their success as farmers, even given all of the struggles with farming that are a part of the process, such as natural disasters, damaging insect infiltrations, economic instabilities and other challenges, the farming arena would look different and we would be much further along in addressing food access for all communities. And yet, the current movement offers opportunities for more creative thinking and inclusive approaches to agriculture that allow for greater participation across communities. There has even been an adjustment in the manner in which farmers are counted in the Census of Agriculture in order to capture the greater participation. In fact, some of the initiatives that were created as a result of historical struggles, such as farming cooperatives and collectives, have actually seen a recent resurgence, largely due to the greater attention that is being paid to the importance of nutrition, along with exercise, as a means to address rising healthcare cost from a preventative standpoint.

The experiences of these four groups Blacks, Native Americans, Latinos and women, when it comes to farming, speak volumes in relation to issues of power and justice. This is because their relationship to land access and land ownership and usage as a vehicle for sustainability for both individuals and communities has always involved a struggle in this country and the outcomes have not been favorable to any of these groups without them having to take simultaneous activist and legal measures. Historically, the United States was founded and further developed with policies that deliberately marginalized each of these groups so that their ability to seek equal opportunity was diminished and, therefore, the fact that this struggle continues to this very day is one that really should not be surprising but instead looked at as a result of the manner in which this country has been structured.

Looking simply at voting, for example, women did not secure the right to vote until 1920 and for African-Americans, it took the Voting Rights Act of 1965 to add some additional protections to voting restrictions they faced at the county level. Today there is still a lot of instances of voter suppression taking place, especially where there are predominant minority populations. Voting is the very basic access to participation in a democratic society, and local elections have determined control of the distribution of federal dollars for farmers. County-level tax officials such as assessors and collectors, also elected, have additionally been critical to the ability to hold onto land. So not being able to participate in these elections, national and local, has been a serious problem.

All of the groups represented by these settlements have deep connections to land and even more particularly food sovereignty and access that provide context to the discriminatory practices they faced. There have also been many struggles associated with the intergenerational transfer of land, even with a tribal nation context that determine decision-making mechanisms, with many local policies making it difficult to both account for and hold on to the land within families. Looking at the Census of Agriculture from the time period of its compilation to the present will give some insight into the diminishing capacity of each of these groups to sustain their individual and collective farms, keeping in mind their seeming choice of deliberate limited participation in this particular census. Various farm bills will also provide some information as to where the priorities of the U.S.D.A. shifted resources away from these marginalized farming groups in the agricultural arena but may have focused some resources of addressing food insecurity in both urban and rural areas that may have ironically targeted some of these same groups.

This book will consider how the need for these other governmental policies related to food access came as a result of the marginalization and oppression experienced by these particular groups in their attempts to sustain farming practices. While the data can't tell us the full extent of the damage sustained by discriminatory practices throughout history, it is the stories that were told in these contemporary settlements that provide a true picture of what has transpired and, more importantly, what needs to be changed. Some debilitating aspects to farming are otherwise cast in the common struggles it takes to be a successful farmer – the market, soil cultivation, uncertain crop yield, natural disasters such as droughts and hurricanes, insect infestations, etc. Then there are some structural impediments – particularly but not solely with African-American farmers, such as heir property rights, which made it hard to determine who should have access to some of the Pigford and Garcia settlements and has also shaped other underserved groups. However, this can also be included in discussions of discriminatory practices of the U.S.D.A. where many farmers who are part of these marginalized groups were not given the same level of recovery support from natural disasters as traditional white male farmers who are specifically mentioned in the Garcia case. There have been some significant changes in outreach services of the U.S.D.A., including a key component of how the Farm Service Agency county committees are structured, which have mitigated future challenges in the area of recovery relief. Time will be the determining factor of success of these political access changes as the Census of Agriculture might indicate a reversal in the land loss suffered by Black and other farmers historically.

Understanding the history of the United States and how its foundation was built on agriculture, land usage and access to farming is critical. This includes aspects of the reservation and the land removal experience of Native Americans, the struggles of Blacks in slavery and post-emancipation Reconstruction policies, throughout the heightened sharecropping era, the necessity and effects of the "Great Migration," to present and Hispanic/Latinos as both farm worker labor and land owners. Additionally, the marginalization of women to limited roles in farming as "wives," of farmers is related to the challenges they have experienced as principal farmers and operators. Information regarding the location and practices of the larger numbers of farms – operators vs. principal operators – will be provided, as well as the geographical location of these settlements which run the gamut across the United States. Regional distinctions of what is farmed, by whom and where the profits lie today in agriculture will be vital information.

Central to this book is the transformation of the agriculture arena which includes efforts of the U.S.D.A., towards more equitable and inclusive access. A part of this transformation has been imaginative and innovative practices that have influenced a greater access to fresh fruits and vegetables through the farm-to-table movement, slow food movement, urban community gardens, Community Support Agriculture efforts and the increasing number of farmers' markets. As stated previously, it is my argument that many of these marginalized groups have largely been at the forefront of these progressive practices that have been instrumental in addressing food deserts and insecurity.

The main point of this book is that changes in the agricultural arena, particularly when it comes to farming are part of movements for social justice – that of a focus of farming justice as part of the food justice movement. Following is a description of each chapter and how connections will be made with regard to each of the four settlements in the context of seeking a more just food system.

Chapter 2 will provide a historical contextualized overview of the complex structure of the U.S.D.A. at the national level in relation to their local agencies and help explain how this type of federalist structure allowed for these discriminatory practices to take place. The focus of this chapter explores the question of land acquisition and power, but also community sustainability. It is important to understand the beginning establishment of the U.S.D.A. and its subsequent changes through various administrations in relation to the specific farming groups that are the focus of this book. This will additionally provide a societal context for understanding the adversity that these groups experienced as a result. For example, at its genesis, what was the connection of the U.S.D.A. to national efforts of land accumulation and expansion that ran up against sovereign indigenous nations? Where was the U.S.D.A. situated during the Reconstruction period when certain promises toward self-sufficiency were given to newly freed Black slaves? And how did this history situate our contemporary inequities and injustices? What about the position of the U.S.D.A. during the women's suffrage movement that initially included a focus on the economic independence of women that was often stripped away through the marriage institution?

Moving to the time period of these cases, largely the 1980s, federal oversight of local agencies became much more limited and actually non-existent due to a change in the United States Administration with the election of President Ronald Reagan who leaned more toward state's rights and less federal influence and intervention. Here is a clear example of our federalist system in action where there is such an extensive connection between federal funding and local government policy implementation, and yet when there is no clear and ongoing oversight, problems can ensue. Additionally, the priorities of the administrations that governed the time period concerning all of these settlements will provide important information and partially explain some of the structural impediments in place. Up until recently, the U.S.D.A. was one of the largest federal agencies and so the priorities of the Executive Branch to a certain extent were made clear through the focus of the U.S.D.A., respectively. However, with the establishment of the Department of Homeland Security after the tragedy of what is commonly referred to as the terrorist act of 9/11, the work of the U.S.D.A. has also become further entangled with a focus on food security with regard to their work in all matters related to agriculture and farming.

Two other areas related to the U.S.D.A. and important to understanding the context of these settlements include the Census of Agriculture, which takes place every five years and also the Farm Bills which are often just as contentious as the overall national budget discussions and provide evidence of the priorities of the presidential administration, which changes with each change of administration. This chapter will give a macro-view level of significant aspects of the U.S.D.A. that created the environment and more specifically policies that allowed for all of the groups represented in this text to suffer discrimination.

Chapter 3 follows with a more in-depth discussion of the federalist structure that governs the U.S.D.A. and how the system operates at the local level. At the local level, there is a considerable amount of community control over federal funding and support for farmers and this is where many of the disparities occurred. It was at this point where a significant portion of the discrimination against these farming groups took place and because that structure is so embedded in the culture of the respective communities this is difficult to change, though some improvements have taken place and will be discussed in both this chapter and in the chapters relative to the specific cases. Included in this chapter will be a discussion of what efforts have been made, thus, far in response to the settlements and whether or not these changes are sustainable and effective to date.

The current policies and practices surrounding the food justice movement overall and the relationship of the U.S.D.A. to these efforts will be discussed because many of these efforts rely upon funding and information from the U.S.D.A. for success, as determined by more inclusive access to healthy food. Here we can make some determination as to whether or not the federal government currently serves as an impediment or as new support, in a form of redress to these relatively new and innovative efforts to address issues of food access and/or food sovereignty.

Chapter 4 explains the history and struggle of Black farmers which led to both of the Pigford settlements, known as Pigford I and Pigford II. Some specific and representative examples of Black farmers and Black farming collectives will be offered to give an intimate perspective of what they experienced given the historical context. It is also important to consider the complexity of the Black farmer experience in terms of heir property which served as a barrier to the process in relation to the maintaining of land ownership. In an effort to hold on to their land through multiple generations, Black families would often distribute various acreage to several family members to both share the wealth but also as a mechanism to protect their property from untoward seizure from government authorities. It is this practice of heir property that the government policies were often in conflict with and made no adjustment for when it came to access to support for farming efforts.

Because this settlement served as a catalyst to the other alleged discrimination cases, the activism that led to this settlement will also be described. It is important to recognize that even after an initial agreement was reached, there was a struggle with the settlement process, which actually involved the U.S.D.o.j. that further demonstrated the entrenched structural problems that needed to be changed. This is why there were two settlements under the Pigford banner. As this chapter will focus specifically on Black farmers, there will be attention drawn to what is taking place today within the context of a farming and food justice movement and efforts to build a sustainable agricultural arena that supports Black farmers. One interesting response of some of the newer Black farmers has been the creation of farm collectives and cooperatives, which does align with a historical tradition of coalescing Black community engagement and support as a means of survival.

Chapter 5 looks at the settlement of Native American Farmers, also known as Keepseagle v. Vilsack. This was the settlement that initially followed that of Black farmers, however, it was different because of the historical relationship between Indigenous nations and the U.S. government that defines both the process and the outcome, and also it involved the issue of water rights. What is similar is that this settlement was challenged in a similar fashion that delayed its processing and even ended up at the level of the Supreme Court, just to be finally settled in the fall of 2018. While the numerous indigenous nation agreements that were determined historically are all different, what is at the heart of this settlement is a more complex issue of tribal sovereignty, not just in government-to-government relations but in the ability of Native American farmers to provide a sustainable living for their community on their land which also bears complex ownership status due to the role of the tribal government. Because of this relationship, the efforts of these farmers today, in terms of a social justice context, are somewhat different. There has also been a national commission, whose efforts will be detailed, established post the Keepseagle settlement to address the ongoing relationship between Native American farmers and the federal government. The work of this commission will be discussed looking especially at the issue of food and water sovereignty that is so central to this case.

Chapter 6 will look at both settlements of Hispanic and women farmers, known as Garcia v. Vilsack and Love v. Vilsack. The discussion of both of these settlements in one chapter is not meant to give the impression that they were not just as significant as the Pigford or Keepsagle cases but is for the reader to understand that while we are talking about two different farmer groups, one that also crosses racial lines, women farmers, the U.S.D.A. handled the finalization of these settlements simultaneously. Additionally, these two groups collectively challenged their final settlement in comparison to that of Black farmers on the basis of fairness. A prominent focus of this chapter, as it looks to what is taking place as a result of these cases, will be the creative and innovative work of women farmers as individuals and collectives, including their extensive network of support that has been ongoing and successful in helping to support more newer women farmers and the specific support they have received from the U.S.D.A. A couple of organizations that are devoted to women in farming, such as the Women, Food and Agriculture Network, the only social justice-focused women farmer organization and the broader field of agriculture will demonstrate how resilient these women farmers have been post the Love settlement. Additionally, this chapter will look at the increasing level of participation of women in federal and local agencies of the U.S.D.A., as well as running for political office and try to determine what differences this has made for women farmers to date. Because women farmers are a substantial part of the global food system, as often primary food producers, international efforts will be mentioned in this chapter, as well as in a latter chapter of the book.

Chapter 7 will return to the contemporary efforts of the U.S.D.A. but also look more broadly at the food justice arena and the U.S.D.A. participation within that arena. The changing structure, outreach efforts, and the shift in the priorities of the U.S.D.A. will be offered and compared to how food justice activists see them and whether or not these same policies are acceptable and/or making a difference. This chapter will also highlight the activism central to the Food Justice Movement that is essentially keeping the U.S.D.A. engaged, regardless of the administration governing the agency. Data from the Census of Agriculture released in April 2019 will help determine whether some significant differences have been made since the settlements of all four cases.

Chapter 8, the concluding chapter will return to the discussion of the struggles of Black, Native American, Hispanic and women farmers to receive recompense from the U.S.D.A. and the connections of their struggles to current food justice efforts. Due to the timing of the writing of this book where the United States is currently in the midst of trying to combat the COVID19 pandemic and the associated impediments especially in relation to food supply and access, there will be a brief discussion regarding the heightened significance of localized food sustainability efforts. It is apparent that with the disruption of the pandemic on both the global and national food supply more people have become aware of what is the availability of locally grown food and more importantly the increased emphasis on food safety. This chapter will look to determine a connection between activist and advocacy efforts in the area of food insecurity, and the similarly focused efforts of the U.S.D.A. Does the future of the agricultural arena look more inclusive and is it a more sustainable approach to nationwide food insecurity challenges? This can be determined by looking both backward and forward to see if the lessons of the past have indeed shaped how the United States moves forward from a farming standpoint.

References

About the U.S. Department of Agriculture|USDA. (n.d.). Retrieved February 1, 2019, from https://www.usda.gov/our-agency/about-usda.

Alexander, K., American Law Division; Ross W. Gorte, Resources, Science, and Industry Division. (2007). Federal land ownership: constitutional authority and the history of acquisition, disposal, and retention, [i]-12.

Bigelow, D., Borchers, A., & Hubbs, T. (2016). *U.S. farmland ownership, tenure, and transfer.* USDA Economic Research Service.

Boyd, J. (1995). History—National Black Farmers Association. http://www.nationalblackfarmersassociation.org/.

Feder, J. & Cowan, T.. (2013). *Garcia v. Vilsack: A policy and legal analysis of a USDA discrimination case.* Congressional Research Service.

Food Sovereignty and the Rights of Indigenous Peoples. (2013). International Indian Treaty Council.

Gottlieb, R. & Joshi, A. (2010). *Food justice.* Cambridge, MA: MIT Press.

Holt-Gimenez, E. (2017). *A foodie's guide to capitalism: Understanding the political economy of what we eat.* New York, NY, USA: Monthly Review Press.

MacDonald, J. M., Hoppe, R. A., & Newton, D. (n.d.). *Three decades of consolidation in U.S. agriculture.* Retrieved February 11, 2019, from http://www.ers.usda.gov/publications/pub-details/?pubid=88056.

Nishime, L. & Williams, K. D. H (eds). (2018). *Racial ecologies.* Seattle, WA: University of Washington Press.

Rodriguez, S (ed.). (2018). *Food justice: A primer.* Sanctuary Publishers.

Taylor, D. E. (2018). Black farmers in the USA and Michigan: Longevity, empowerment, and food sovereignty. *Journal of African American Studies.* 22, 49–76. https://doi.org/10.1007/s12111-018-9394-8.

Tribal Food Sovereignty definition. (n.d.). [Wellforculture.com].

Universal Declaration of Human Rights. (1948). United Nations General Assembly. https://www.un.org/en/universal-declaration-human-rights/.

USDA - National Agricultural Statistics Service—Census of Agriculture. (n.d.). Retrieved February 19, 2019, from https://www.nass.usda.gov/AgCensus/.

USDA Defines Food Deserts | American Nutrition Association. (n.d.). Retrieved February 11, 2019, from http://americannutritionassociation.org/newsletter/usda-defines-food-deserts.

USDA ERS - Food Access Research Atlas. (n.d.). Retrieved February 11, 2019, from https://www.ers.usda.gov/data/fooddesert/.

2 The politics of the U.S.D.A., the Census of Agriculture and Farm Bills

This chapter provides a historical and politically contextualized overview of the complex structure of the United States Department of Agriculture (U.S.D.A.) at the national level and its relationship to and oversight of the work of the local U.S.D.A. agencies. The system of federalism that shapes the national U.S. government in relationship to the state and local governments means that many federal policies are implemented by passing funds to the state and local agencies. The federal government implements policies in relation to and through state and local governments. However, the lack of sustained oversight within the U.S.D.A. between the federal headquarters and the local agencies that were largely governed by elected county committees helps to understand how ongoing alleged discriminatory practices were allowed to take place for so many years. Some of this oversight was handled by a civil rights division within the U.S.D.A., which was intended to ensure that those groups (which are the focus of this book) protected by the Civil Rights Act of 1964 had equal access to resources and support. Unfortunately, this division was eliminated in 1983 by President Reagan who eliminated the U.S.D.A. Office of Civil Rights (OCR) when he was pushing through budget cuts (Krom, 2010).

There is clearly an understanding of the disparities that exist in agricultural policies that have race and gender at the center but it has taken both the efforts of farming and food justice activists, as well as, committed agricultural policy experts more situated in the United States government system, specifically the U.S.D.A., to collectively work toward making real changes. In 1997, then U.S.D.A. Agriculture Secretary Dan Glickman put together a Civil Rights Action Team composed of senior U.S.D.A. officials to determine if both internal and external civil rights complaints were being addressed in a timely manner by the fairly newly created O.C.R. which reported to the Assistant Secretary for Administration. The report titled "USDA – Problems Continue to

Hinder the Timely Processing of Discrimination Complaints," found the following:

1 U.S.D.A. lacked the organizational structure to support an effective civil rights program;
2 U.S.D.A.'s process for resolving discrimination complaints about the delivery of program benefits and services (program complaints) was a failure and;
3 U.S.D.A.'s system for addressing complaints of employment discrimination (employment complaints) was untimely and unresponsive (*U.S. Department of Agriculture*, 2012).

The team realized that there was not enough staffing or managerial expertise and they also discovered that the O.C.R. did not have the best working relationships with other U.S.D.A. entities.

In a report issued in November 2010, U.S.D.A. Secretary of Agriculture Tom Vilsack discussed his cultural transformation initiative which was critically important to moving the agency toward more inclusive practices both internally and externally. He stated,

> In terms of personnel, it looks like a modern-day snapshot of America. It includes a workforce that reflects the demographic make-up of the citizens we serve. In terms of organization, Cultural Transformation means adapting to the technological advancements of social media and recognizing their cultural implications to our organizational performance. Our ability to effectively respond to our customer needs requires us to incorporate the most recent communication and information methods into our work culture. That doesn't mean abandoning customers who aren't online or tech-savvy, but it does mean recognizing that our work processes must be compatible with all our customers.
>
> (Vilsack, 2010)

Without the acknowledgment of that necessary cultural shift, some of the changes needed at the U.S.D.A. to make their policies and programs more accessible may never have moved forward. It was important to have that statement from the top agency official clearly spelling out the goal to shift the culture of the institution and make it more representative and accessible.

Though the alleged discrimination settlements that are at the heart of this book, namely Pigford, Keepseagle, Garcia and Love covered a time frame within the 1980s to the 1990s, they were reflective of a longer period of struggles for these farming groups not just in connection to their relationships with the U.S.D.A. but as relatively small family farms struggling to survive through all types of economic, climate and environmental conditions that have tremendous effects on their viability. Because for many reasons they were already in vulnerable positions, access to the level of resources needed to thrive was tenuous at best, sometimes due to the way the policies and programs were designed and sometimes simply due to lack of awareness. When taken into consideration some of the survival practices of these groups created as shields from oppressive systemic conditions, it is easy to surmise how corrective measures that were taken up by the U.S.D.A. to address some of the racial and gender disparities with these particular farming groups may have fallen short. Ironically, some of these practices also shape the work of food justice activists to address disparities that required a more urgent response than navigating the red tape of government policies would allow.

One example of a survival mechanism used to hold onto acquired land that is out of alignment with government policies of access to resources is the practice of heir property. This was one practice that African-Americans, Latinos and Native Americans used to ensure maximum benefit of land ownership with distribution to family members which unfortunately negated access to U.S.D.A. farming support. The constraints of those policies prescribed that land must be owned by only one person, and thus cast in a model of a traditional family structure which often ran counter to a more extended family make-up. This will be discussed in greater detail in subsequent chapters.

Another example is through the historic complex relationship of sovereign tribal governments and the U.S. federal government in the form of various treaties and additionally in the manner in which these treaties were governed through the Bureau of Indian Affairs with the Department of Interior. Then there are the marginalizing policies set in traditional patriarchal values that undermined the ability for women to hold onto their own land and farms or even be counted as partial owners. Looking at all of these distinct groups, their collective farm experiences fit within the complex growth and development of American society and the ongoing changes that shape the field of agriculture specifically. Much of their struggles and the inability to convince the government to intervene have been lost in the rugged white male individualism narrative of the United States, which is often framed as Dave Mason suggests in his book "The End of the American Century,"

> When things are going well, Americans attribute it to their own personal achievements and when things are going badly, they assume that they themselves (or the other victims) are to blame. Americans tend not to blame the system. Of course, American individualism is a central element of the national self-image and a source of much national pride. It also sets the United States off, for better or worse, from most other countries and has become a component of American "exceptionalism."

> (Mason, 2009)

Their collective struggles have also been lost in the American ideological underpinnings of agrarianism which Deborah P. Dixon and Holly Hapke noted Thomas Jefferson as describing it in the following ways: 1) belief in the independence and virtue of the yeoman farmer; 2) the concept of private property as a right; 3) land ownership without restrictions on use or disposition; 4) the use of land as a safety valve to ensure justice in the city and; 5) the conviction that with hard work, anyone could thrive in farming. And they further state Jefferson as believing "landowning farmers were pivotal to the creation and maintenance of a democratic society" (Dixon & Hapke, 2003). There has been an unrelenting disconnect between American stated values and the universal application of these values to all people.

One key focus is the question of land acquisition, power and community sustainability with regard to how certain groups struggled through adversity to be self-sufficient through farming. Some of these struggles occurred just in the natural course of challenges faced by farmers but some of their struggles were due to the inability of these particular farming groups to access support from the U.S.D.A., particularly at the local level. How and why this happened can be understood through knowledge of the historical political and economic context that led to the establishment of the U.S.D.A. and its subsequent changes and priorities through various administrations. Furthermore, because a critical aspect of the challenges these farming groups faced actually occurred at local levels, the discussions of the specific alleged discrimination cases will provide more of a United States regional understanding that is related to where the development of this country shaped migratory and settlement patterns of groups. Generally stated, this focus is about the experiences of Black farmers in the South, Native

American farmers in the Northeast, women farmers in the Midwest and Latino/Hispanic farmers in Texas and the Southwest. These farming groups are not limited or isolated to the regions mentioned here, it is just the predominance of their locations which was exemplified in their specific settlements. Thus, looking at these conditions through a critical racial and feminist theoretical lens, we see the plantation economy, westward land expansion, the growth and development of American statehood and the border expansions that inform both the historical and contemporary experiences of the collective marginalized farming groups at the center of this book. While specifically, the racial and ethnic groups involved in these settlements occupied those aforementioned regional areas in significant numbers, they were not fully a part of the power structure with equitable access to resources, such as land, with regard to U.S. government development, and therefore they suffered disproportionately.

The question of land accessibility and ownership is central to the issue of farming and food sovereignty and the involvement of the United States government in creating policies in this area provides a systemic understanding of how disparities originated. According to the U.S.D.A. report, U.S. Farmland, Ownership, Tenure and Transfer, there have been ongoing shifts in the distribution of owned acres vs. rented land, most dramatically between the period following the Great Depression in the 1930 and post one of the most contemporary recession period around 2012. "Between 1935 and 2012, the percentage of acres in full-owner operations, where the operator owns all of the land on the farm, remained relatively stable at 37 percent in 1935, 34 percent in 1954, and 37 percent in 2012" (U.S. Census Bureau, 1935, 1954; USDA-NASS, 2012). The most significant historical change in tenure derived from full-tenant operations (32 percent of the acreage in 1935 and 10 percent in 2012) shifting to part-owner operations – 25 percent in 1935 and 54 percent in 2012. (Bigelow, Borchers, & Hubbs, 2016)

The historical context explains the approach of many of these small farmers involved in the food justice movement today which relies heavily on survival strategies that were developed in the midst of systemic and environmental impediments. Today, the field of agriculture has evolved and there are deliberate efforts on the part of the U.S.D.A. to make their resources and support more accessible. However, because there is such a long history of inequity that exists outside of the years that the settlement covers, given these roots, some necessary systemic changes may never be fully addressed. Some of the institutional changes at the U.S.D.A. came as a result of the settlements, some changes as a result in the changing of administrations and some of the policies of today are simply determined by both local and global economic changes happening in the field of agriculture – such as the advancement of technology and the use of social media as a marketing tool.

President Abraham Lincoln established the U.S.D.A. in 1862, which he called "The People's Department." This agency was established less than a year before he signed the Emancipation Proclamation, which effectively declared all slaves free who resided in territory that was in rebellion against the federal government, contingent upon the Union winning the war. This covered approximately 3.5 million of the 4 million who were known to be enslaved. In her book on the Emancipation Proclamation, author Judy Dodge Cummings points out that this was a temporary war measure that ultimately led to permanent freedom because President Abraham Lincoln used the Constitutional process to get it passed. As she states, it "went farther than any other law in that it recognized African-Americans as people not property" (Cummings, Judy Dodge, 2016). However, Cummings goes on to note that the protections of the Emancipation Proclamation did not last beyond a decade because many states instituted Black Codes that prevented Black people from doing things like voting, serving on juries, testifying against whites, owning guns and a number of states barred Black people from owning land. (Cummings, 2016). Some lingering aspects of these Black codes, where laws and policies are developed at the state and local level still remain challenges for Black people and other racial and ethnic minorities today.

The initial establishment of the U.S.D.A. was to provide more education and support for the nation's farmers. The intentions of President Lincoln in establishing the U.S.D.A. encompassed a set of values centered on the best way to sustain a rapidly growing country while maximizing the use of land acquisition. However, as noted on the U.S.D.A. website and included as part of its 150th Anniversary Celebration documents, the only extensive speech Lincoln gave regarding agriculture occurred in Milwaukee, Wisconsin on September 30, 1859 (*USDA History Collection Introduction/Index | Special Collections*, n.d.). In this speech, Lincoln acknowledged that he was not the most expert when it came to the agricultural arena, but that he did have his own personal connections to farming. At the time of his speech, however, he was speaking more from his position as a politician. Toward the beginning of his address at the Agricultural Fair, President Lincoln stated the following before offering some detailed suggestions on agricultural development:

> One feature, I believe, of every Fair, is a regular Address. The Agricultural Society of the young, prosperous, and soon to be, great State of Wisconsin, has done me the high honor of selecting me to make that address upon this occasion–an honor for which I make my profound and grateful acknowledgment. I presume I am not expected to employ the time assigned me in the mere flattery of the farmers, as a class. My opinion of them is that, in proportion to numbers, they are neither better nor worse than other people. In the nature of things they are more numerous than any other class; and I believe there really are more attempts at flattering them than any other; the reason of which I cannot perceive, unless it be that they can cast more votes than any other. On reflection, I am not quite sure that there is not cause of suspicion against you, in selecting me, in some sort a politician, and in no sort farmer, to address you. But farmers, being the most numerous class, it follows that their interest is the largest interest. It also follows that interest is most worthy of all to be cherished and cultivated–that if there be inevitable conflict between that interest and any other, that other should yield. Again, I suppose it is not expected of me to impart to you much specific information on Agriculture. You have no reason to believe, and do not believe, that I possess it–if that were what you seek in this address, any one of your own number, or class, would be more able to furnish it. You, perhaps, do expect me to give some general interest to the occasion; and to make some general suggestions, on practical matters. I shall attempt nothing more. And in such suggestions by me, quite likely very little will be new to you, and a large part of the rest possibly already known to be erroneous.
>
> (Lincoln's Milwaukee Speech | National Agricultural Library, n.d.)

While President Lincoln distanced himself from identifying directly with the farming class, as a politician he recognized his power in numbers as an electorate to which he needed to be responsive. Two priorities that were a part of the Republican Party Platform in 1860, where he secured the nomination for President included 1) a demand for a homestead measure which certainly caused a lot of conflict with indigenous nations and 2) he advocated for federal aid for the construction of a railroad to the Pacific Ocean (*USDA History Collection Introduction/Index | Special Collections*, n.d.)

The creation of the U.S.D.A. led to the earliest considerations of what we call our "food system, rooted in a racialized political structure because it deliberately did not include a collaborative and mutually respectful relationship with the indigenous nations and additionally efforts were not inclusive of the soon to be emancipated slaves. As the country grew, and the need for food grew, respectively, all manners of human sustenance shifted and yet the racial and sovereign disparities were just exacerbated. Where the priorities of sustainability were apparent, there also existed the conflicting priorities of growing the economy to benefit those who had the money to invest in this burgeoning capitalist system, to the detriment of those who were relegated to menial labor conditions including slavery, sharecropping, tenant farming and the like without much if any personal or economic benefit from their own labor. And ultimately land as directly associated with power and wealth explained the conflated relationship between agriculture, power and status, and additional policies were put in place to keep minority and ethnic groups marginalized. Some policy outcomes were unintended but they had the same effect. Eventually, the ability to participate in the political decision-making process did shift from the requirement of land ownership but participation still remained within an elite-strata defined by wealth, among other factors such as gender (male) and being white. However, to a certain extent land remained one determining factor in affecting power and being independent as evidenced by the myriad of laws and policies that made it easy to take land away from the specific groups who are the focus of this text. These laws and policies prevented not only the ability for these groups to progress economically but also prevented them from a collective acquisition of power and even stability.

The alleged discrimination faced by Black, Native American, Hispanic/Latino(a) and women farmers can be explained by the historical circumstance of when the U.S.D.A. was established – during the midst of the U.S. Civil War. At that time, the country was wrestling with the conflict situated in the succession of southern states from the national government largely due to their insistence on preserving the institution of slavery as central to their economic vitality. This is not to suggest that there were no businesses in the northern states which also benefitted from the institution of slavery. The beneficial relationship of northern states to slavery was indirect because many of them had already passed anti-slavery laws. Under the larger concern of "state's rights," the reality was that the majority of the country was agricultural-based, the population was growing and there were conscious efforts toward land expansion. Therefore, national resources were determined to be necessary for sustainability and progress, which was a priority of the national government prior to the start of the U.S. Civil War.

As detailed by the history on the U.S.D.A. website, President Lincoln, in his annual message to Congress on December 3, 1861 stated *"Agriculture, confessedly the largest interest of the nation, has not a department nor a bureau, but a clerkship only, assigned to it in the Government. While it is fortunate that this great interest is so independent in its nature as to not have demanded and extorted more from the Government, I respectfully ask Congress to consider whether something more cannot be given voluntarily with general advantage.... While I make no suggestions as to details, I venture the opinion that an agricultural and statistical bureau might profitably be organized."* (USDA History Collection Introduction/Index | Special Collections, n.d.)

He signed that law establishing the U.S.D.A. on May 15, 1862, and on May 20, 1862, President Lincoln signed the Homestead Act which "provided for giving 160 acres of the public domain to any American or prospective citizen who was the head of a family or over 21 years of age. Title to the land was issued after the settler had resided on it for five years and made improvements on it. The settler could also gain title by residing on the claim for six months, improving the land, and paying $1.25 per acre. (USDA.gov) Roxanne Dunbar-Ortiz points out

"During the Civil War, with the southern states unrepresented, Congress at Lincoln's behest passed the Homestead Act in 1862, as well as the Morrill Act, the latter transferring large tracts of Indigenous land to the states to establish land grant universities. The Pacific Railroad Act provided companies with nearly two hundred million acres of Indigenous land. With these land grabs, the US government broke multiple treaties with Indigenous nations. Most of the western territories, including Colorado, North and South Dakota, Montana, Washington, Idaho, Wyoming, Utah, New Mexico and Arizona, were delayed in achieving statehood, because Indigenous nations resisted appropriation of their lands and outnumbered settlers.... As industrialization quickened land as a commodity, "real estate," remained the basis to the US economy and capital accumulation. The federal land grants to the railroad barons, carved out of Indigenous territories, were not limited to the width of the railroad tracks, but rather formed a checkerboard of square-mile sections stretching for dozens of miles on both sides of the right of way.

(Dunbar-Ortiz, 2014)

Later that year, on July 2, 1862, President Lincoln signed the Morrill Land Grant College Act, which donated public land to the States for colleges of agriculture and the mechanical arts, agreement that was accepted by each of the states at the time to establish one or two institutions that fit that description. Between 1886 and 1932, there were several types of initiatives to promote more independent farms by black tenants and sharecroppers. The most famous and durable achievements were in agricultural education. The Second Morrill Act of 1890 established state agricultural colleges for black students and Booker T. Washington (1856–1915) emerged as a leading public figure in promoting education and farm improvement for Black people. But three other initiatives of this period applied directly to the development of independent farming: (1) organization of cooperatives for farmers and other community services, (2) projects for land purchase and resale to small farmers, and (3) farm diversification and self-sufficiency.

It is important to note that no African-Americans were allowed to attend the original land-grant institutions at the time, which lead the way for historically black colleges and universities with similar interests, like Tuskegee University to be founded. In his autobiography, Up From Slavery, Booker T. Washington states,

> "We found that most of our students came from the country districts, where agriculture in some form or other was the main dependence of the people. We learned that about eighty-five per cent of the coloured people in the Gulf states depended upon agriculture for their living. Since this was true, we wanted to be careful not to educate our students out of sympathy with agricultural life, so that they would be attracted from the country to the cities, and yield to the temptation of trying to live by their wits. We wanted to give them such an education as would fit a large proportion of them to be teachers, and at the same time cause them to return to the plantation districts and show the people there how to put new energy and new ideas into farming, as well as into the intellectual and moral and religious life of the people." (p. 97)
>
> (Washington, Dubois, & Johson, 1901)

Though Lincoln's position on post-slavery emancipation was that the former slaves would willingly go to somewhere like Panama or back to the African-continent, the details of that policy were not fully explored prior to his assassination. In fact, Ibram X. Kendi noted,

"If I could save the Union without freeing any slave I would do it, and if I could save it by freeing all the slaves I would do that," President Abraham Lincoln wrote on August 20, 1862. "What I do about slavery, and the colored race, I do because I believe it helps to save the Union." On January 1, 1863, Lincoln signed the Emancipation Proclamation as a 'necessary war measure.'

(Kendi, 2019)

There is no doubt that while there was a goal of land expansion as a means to grow the country, policies that involved federal land distribution, for the explicit purpose of farming, were geared toward recent white European immigrants and did not include the newly emancipated slaves of African descent at the end of the over 250-year period of enslavement for them. In fact, any Reconstruction policies initiated to help the newly emancipated slaves obtain land were very short-lived.

I attended Auburn University, the first land-grant institution in the South, and from this experience witnessed firsthand the importance of the present-day connection of these types of institutions to rural community development generally and agriculture in particular. In connection to my graduate studies at Auburn, I spent quite a bit of time in Wilcox County, which is a part of the Black Belt area of Alabama and spent a considerable amount of time in the Auburn Extension Office as they were provided a number of services in the community related to farming information and food nutrition in general. Auburn University history is indicated as following on their website:

"The university began, though, as the small, more humble East Alabama Male College, which was chartered in 1856 and opened its doors in 1859 as a private liberal arts institution. From 1861 to 1866 the college was closed because of the Civil War. The college had begun an affiliation with the Methodist Church before the war. Due to dire financial straits, the church transferred legal control of the institution to the state in 1872, making it the first land-grant college in the South to be established separate from the state university. It thus became the Agricultural and Mechanical College of Alabama. A land-grant college or university is an institution that has been designated by its state legislature or Congress to receive the benefits of the Morrill Acts of 1862 and 1890. The original mission of these institutions, as set forth in the first Morrill Act, was to teach agriculture, military tactics, and the mechanical arts as well as classical studies so that members of the working classes could obtain a liberal, practical education."

(The History, n.d.)

Because Alabama is still today noted as primarily an agriculturally based state with significant focus and resources tied to farming and the economic development of rural communities it was a good place to study and get a firsthand perspective on the relationship between these types of academic institutions and the U.S. government at all levels, federal, state and local, which are partnered through agricultural-extension efforts.

Today, there are nearly agricultural extension offices in most of the 3000 counties of the U.S., according to the U.S.D.A. website. These offices were formally established in 1914 by the Smith Lever Act, which secured the partnership of the U.S.D.A. with state land-grant universities for the purpose of conducting research and providing education in agriculture. While this book is focused most specifically on the financial support and access of marginalized farmers to U.S.D.A. support, it is important to consider what type of access was and is provided through the agricultural extension offices, which today are on the forefront of supporting both rural and urban farming efforts across the country (*About the U.S. Department of Agriculture | USDA*, n.d.)

The priorities of the U.S.D.A. shifted continuously, depending on the administration and more directly the person who was selected to lead the agency. Because of the critical reliance on support provided by the U.S.D.A., what is now known as the Secretary of Agriculture is a highly politicized position with those selected today sometimes seen as having a predisposition toward large agribusiness corporations and the movement into the international trade arena, and not as accessible to smaller family-oriented farms and farming collective groups. This has often complicated the relationship between policy priorities, especially some that both directly and indirectly affect smaller minority and women farmers and the global reliance upon large U.S.-based agribusiness corporations which have such a tremendous effect on the food supply. Additionally, the work of the U.S.D.A. has always changed with the needs and demands of the U.S., especially during times of war and/or economic strife. In 2012, the U.S.D.A. celebrated its150th Anniversary and noted all of the particular shifts that occurred under various administrations.

The first "Commissioner," of the Department of Agriculture, selected by President Abraham Lincoln was a farmer by the name of Isaac Newton, a farmer who had served as chief of the agricultural section of the Patent Office since August 1861. Newton was born in Burlington County, New Jersey. He grew up on a farm, and after completing his common-school education, became a farmer in Delaware

County, Pennsylvania, near Philadelphia. Newton was a successful, progressive manager, whose farms were regarded as models. He also developed a pioneer dairy lunch in Philadelphia and a select butter trade as outlets for his farm products. Newton sent butter each week to the White House; and he and his family maintained a close friendship with the Lincolns. Subsequently, Lincoln gave him full support in managing the Department (*USDA History Collection Introduction/ Index | Special Collections*, n.d.).

In his first annual report, Newton outlined objectives for the Department. These were: (1) collecting, arranging and publishing statistical and other useful agricultural information; (2) introducing valuable plants and animals; (3) answering inquiries of farmers regarding agriculture; (4) testing agricultural implements; (5) conducting chemical analyses of soils, grains, fruits, plants, vegetables, and manures; (6) establishing a professorship of botany and entomology; and (7) establishing an agricultural library and museum. These objectives were similar to the charges given the department by Congress in its legislation establishing the new agency. A U.S.D.A. video focusing on the priorities of the Secretary of Agriculture specifically notes that while the overcharge of the U.S.D.A. is to speak to the very basic needs of the everyone in the country and, thus, leads it to be less partisan than other government agencies, the political climate of the country has heavily influenced their policies and practices. For example, Secretary Bob Bergland (1977–1981 Ag Secretary) discussed their response after receiving a call from President Jimmy Carter about initiating a grain embargo and how they needed to quickly adjust their practices for grain farmers. He additionally mentions how the U.S. Forest Services, another U.S.D.A. agency had to address some key environmental concerns of the lumber industry, under pressure from the environmental movement at that time. Other important points mentioned include the American Agricultural Movement which he says, "They had a lot of property, but they had a lot of debt." Table 2.1 provides information on the U.S.D.A. Secretaries that served relative to the time period focused on in the alleged discrimination lawsuits at the heart of this book.

There is one particular agency within the U.S.D.A. that directly serves to support farmers, known today as the Farm Services Agency. Looking historically at this organization, the Farm Services Agency can trace its beginnings to 1933, in response to the Great Depression and President Franklin Delano Roosevelt's efforts to address the increasing number of farms that were failing. One example of a community that I am familiar with that benefitted from these efforts at that

Table 2.1 USDA Secretaries 1977–2016 – from service under President Jimmy Carter to service under President Barack Obama

Secretary	Served under	Policy information
Robert Selber Bergland (Minnesota)	Jimmy Carter 1977–1981	20th USDA Secretary - Food and Agricultural act of 1977: Increased price and income supports National Agriculture, research, teaching policy: Made USDA leading agency for AG research and consolidated funds. Farmer strike because commodity prices became lower than price of production
John Rusling Block (Illinois)	Ronald Reagan 1981–1986	21st USDA Secretary - !985 Farm Bill - Food Security Act Abolished USDA Civil Rights office in 1981 Most farm loss since great depression. Colman V. Block= USDA not giving enough notice of debt restriction options. Later Led to Ag Credit Act of 1987
Richard Edmund Lyng (California)	Ronal Reagan 1986–1989	22nd USDA Secretary - Overall goal was to limit government involvement. He a sent memo to all departments that race would be central issue and they will be judged by racial discrimination records
Clayton Keith Yeutter (Nebraska)	George HW Bush 1981–1991	23rd USDA Secretary - Food, Agriculture, Conservation, Trade Act of 1990 - new rural development administration. Administration shaped by decreasing cost of programs, increasing open markets, and concern about effects of pesticides and chemicals
Edward Rell Madigan (Illinois)	George HW 1991–1992	24th USDA Secretary – Businessman from Illinois. Elected as a Republican to the 93rd and to the nine succeeding Congresses (January 3, 1973–March 8, 1991); resigned March 8, 1991 prior to serving as Secretary of Agriculture
Alphonso Michael Epsy (Mississippi)	Bill Clinton 1993–1994	25th and first African American USDA Secretary - Provided relief for farmers after Mississippi river flood in 1993. Opened Markets for farmers internationally through NAFTA. Charged with ethic violations and stepped down.
Daniel Robert Glickman (Kansas)	Bill Clinton 1995–2000	26th USDA Secretary - Freedom to Farm Act of 1996. Reduced premium for crops and increased farmers insurance coverages and established National Drought Policy.

Table 2.1 (Continued)

Secretary	Served under	Policy information
Ann Veneman (California)	George Bush 2001–2005	27th USDA Secretary - Focused on Infrastructure enhancement, conservation of environment, rural communities, and trade expansion through new export markets. Established Leaders of Tomorrow education program and USDA E-Government initiative so programs and services available electronically.
Mike Johanns (Iowa)	George Bush 2005–2007	28th USDA Secretary - Early focus was to reopen beef markets after mad Cow scare Worked on Doha Round to open up international trade - ultimately unsuccessful
Ed Schafer (North Dakota)	George Bush 2008–2009	29th USDA Secretary - Early focus on animal welfare and human food safety. Farm Bill was in negotiations with Congress. Believed there were fraudulent claims in Pigford case against USDA.
Tom Vilsack (Iowa)	Barack Obama 2009–2016	30th USDA Secretary – helped pass Healthy, Hunger Free Kids Act. Focused on improvement of the American food supply and made civil rights a priority. Settled Pigford, Keepseagle, Garcia and Love cases. New trade agreements with Columbia, Panama and South Korea.

time is a community called "Gee's Bend," in Wilcox County, Alabama. Many people have heard of the Gee's Bend Quilting Bee, as it was one of the first Black Southern Women's Cooperative founded in the 1960s. Gee's Bend was named after Joseph Gee, who relocated there from Halifax County in North Carolina in 1816.

He brought 18 enslaved blacks with him and established a cotton plantation. When he died, he left 47 slaves and his estate to two of his nephews, Sterling and Charles Gee. In 1845, the Gee brothers sold the plantation to a relative, Mark H. Pettway, and the Pettway family name remains prominent in Wilcox County. After emancipation, freed blacks who stayed on at the plantation worked as sharecroppers and farmers. The Pettway family held the land until 1895, when they sold it to Adrian Sebastian Van de Graaff, an attorney from Tuscaloosa who operated the plantation as an absentee landowner.

(Stevens, 2007)

In 1937, the owners of the land at that time, the Van de Graff family, sold it to the Farm Security Agency who turned the community into Gee's Bend Farms, Inc, as a cooperative. Today, there is a Gee's Bend Ferry Service that transports citizens and visitors to the community from Camden, the county seat of Wilcox County, Alabama over to Gee's Bend.

Initially called the Farm Security Agency and connected to the Resettlement Administration, "its original mission was to relocate entire farm communities to areas in which it was hoped farming could be carried out more profitably." This project, according to the U.S.D.A. website, turned out to be both controversial and expensive so the Standard Rural Rehabilitation Loan Program was established, "which provided credit, farm and home management planning and technical supervision," serving as a model for the establishment of the Farmers Home Administration." Having met with some success in elevating the incomes of many poor families, in 1946 the Farmers Home Administration consolidated the Farm Security Administration with the Emergency Crop and Loan Feed Division of the Farm Credit Administration (*USDA History Collection Introduction/Index | Special Collections*, n.d.).

During this same time, the Agricultural Adjustment Administration (A.A.A.) was established by Act in 1933. The "Triple A's" purpose was to stabilize farm prices at a level at which farmers could survive which was done through the first federal farm program offering price support loans to farmers to bring about crop reduction. The oversight of this program was done by county committees and is one area where a significant number of alleged discriminations began. Most of these county committees were controlled by white men who deliberately discriminated against minority farmers, not processing loan applications or processing them too slow and then subsequently micro-managing the ability of these minority farmers to use the funding as they saw best (Krom, 2010). There has since been a deliberate effort made on the part of the U.S.D.A. and many county committees to ensure that there is at least some level of minority representation even if the voting process does not meet that objective (*2019 County Committee Elections – Farm Services Agency*, 2019)

In 1994, a reorganization of U.S.D.A. resulted in the Consolidated Farm Service Agency, renamed Farm Service Agency in November 1995. The new F.S.A. encompassed the Agricultural Stabilization and Conservation Service, Federal Crop Insurance Corporation (F.C.I.C.) and the farm credit portion of the Farmers Home Administration.

In May 1996, F.C.I.C. became the Risk Management Agency. Today, F.S.A.'s responsibilities are organized into five areas: Farm Programs, Farm Loans, Commodity Operations, Management and State Operations. The agency continues to provide America's farmers with a strong safety net through the administration of farm commodity programs. F.S.A. also implements ad hoc disaster programs. F.S.A.'s long-standing tradition of conserving the nation's natural resources continues through the Conservation Reserve Program. The agency provides credit to agricultural producers who are unable to receive private, commercial credit. F.S.A. places special emphasis on providing loans to beginning, minority and women farmers and ranchers.

A separate entity known as the Census of Agriculture (Ag Census) which takes place every five years is also significant to understanding some of the difficulties associated with the farming groups that are the focus of this book. The manner in which data has been historically gathered to account for who is part of the farming arena in this country has also been wrought with challenges that have shaped the inability for minority and women farmers to adequately access resources, especially from the U.S.D.A. There are many other agencies that rely upon this data for policy decisions, such as the Economic Research Service or any organization interested in keeping up with the progress of commodity growth in the United States.

The collection of agricultural census data was initially authorized by the United States Constitution in Article 1, Section 2, which required a census of population to be conducted every 10 years to proportionately distribute the representation of each state in the United States House of Representatives. The census of agriculture originated as part of the 1820 decennial census when U.S. marshals began to ask how many people within each household were engaged in agricultural pursuits. In 1840, marshals began using separate census schedules to collect data related to agriculture. The Ag Census details the creation, cultivation and distribution of all homegrown commodities of which this country relies upon for sustenance. It provides geographical information, state-by-state determination of what is being grown and raised, profits associated with these agriculture efforts and even provides demographic information, which is key. The Ag Census also is important as we look across time to measure any changes in determining what commodities are most sustainable and valuable to society. And it is possible that we can take the Ag Census and look at data concerning poverty concentration in this country and determine patterns that need to be addressed from a policy perspective (*2007 Census of Agriculture – History*, 2007).

Table 2.2 Census of Agriculture accounting of specific farming groups

Year	Total farms	Black farms	Black percentages	Hispanic farms	Hispanic percentages	Native American farms	Native American percentages	Women farms	Women percentages
1978	2,257,775	37,351	1.65%	17,476	0.77%	6,889	0.31%	112,799	5.00%
1982	2,240,976	33,250	1.48%	16,183	0.72%	7,211	0.32%	121,599	5.43%
1987	2,087,759	22,954	1.10%	17,476	0.84%	7,134	0.34%	131,641	6.31%
1992	1,925,300	18,816	0.98%	20,956	1.09%	8,346	0.43%	145,156	7.54%
1997	2,215,876	18,451	0.83%	27,717	1.25%	10,638	0.48%	165,102	7.45%
2002									
Principal operators	2,128,982	29,090	1.37%	50,592	2.38%	15,494	0.73%	237,819	11.17%
All operators	*3,115,172*	*37,791*	*1.21%*	*72,349*	*2.32%*	*42,304*	*1.36%*	*847,832*	*27.22%*
2007									
Principal	2,204,792	30,599	1.39%	55,570	2.52%	34,706	1.57%	306,209	13.89%
All	*3,281,534*	*39,697*	*1.21%*	*82,462*	*2.51%*	*55,889*	*1.70%*	*985,192*	*30.02%*
2012									
Principal	2,109,303	33,371	1.58%	67,000	3.18%	37,851	1.79%	288,264	13.67%
All	*3,180,074*	*44,629*	*1.40%*	*99,734*	*3.14%*	*58,475*	*1.84%*	*969,672*	*30.49%*
2017									
Farms	2,042,220	35,470	1.74%	86,278	4.22%	60,083	2.94%	1,139,675	55.81%
Producers	3,399,834	48,697	1.43%	112,451	3.31%	79,198	2.33%	1,227,461	36.10%

Before 2002, the Census of Agriculture collected detailed demographic data on only one operator per farm. Since 2002, the census has taken a more comprehensive approach, counting all operators and collecting detailed demographic information on up to three operators per farm. The principal operator is the person in charge of day-to-day decisions for the farm or ranch. In 2017, the language changed again looking at producers and farmers

As we continue to look at its evolution between 1850 and 1920, it remained connected to the decennial census program, adding economic data in 1925, however, there was a change in 1954. The U.S. Census Bureau decided to conduct the Ag Census instead in years ending in "4" and "9." Moving toward how the Census of Agriculture is conducted today, responsibility for conducting this task went from the U.S. Department of Commerce, Bureau of the Census, to the U.S.D.A., National Agricultural Statistics Service. The agriculture census is the only source of statistics on American agriculture showing comparable data, by county and classifying farms by size, tenure, type of organization, primary occupation, age of operator, market value of agricultural products sold, combined government payments and market value of agricultural products sold and North American Industry Classification System codes.

As we understand the history of the U.S.D.A. and the significance of the data collected through the Census of Agriculture, the recognition of policies and practices of organizations responsible for attention to all farmers and farming groups become apparent, especially post the Pigford, Keepseagle, Garcia and Love settlements where the respective agencies took intentional corrective measures in their outreach efforts. As noted with regard to the 2007 Census,

> Working in partnership with community-based organizations nationwide, NASS (National Agricultural Statistics Service) engaged in an extensive effort to make sure the 2007 Census mailing list included all farms and ranches, regardless of size, location or type of operation. NASS also partnered with these groups to provide hands-on assistance and support to local producers – including non-English speakers – in filling out their Census forms. In addition, NASS targeted its media outreach efforts towards publications and broadcast outlets that reach small, minority and non-English speaking producers. Also, the 2007 Census of Agriculture marked the first time NASS attempted to collect a Census report from individual farm operators on American Indian reservations in all states. In 2002, reservations were counted as a single farm and reservation officials supplied counts of individual operators. NASS will issue a follow-up report about agricultural activity on American Indian reservations in the spring of 2009. NASS is committed to continuing its outreach to all underserved populations.

> (2007 Census of Agriculture – History, 2007)

Today the official description of the Census of Agriculture states the following:

> It provides a detailed picture of U.S. farms and ranches every five years and is the only source of uniform, comprehensive agricultural data for every county or county equivalent in the U.S. Agriculture census data are routinely used by Congress; Federal, State, and local government organizations; the business community; scientific and educational institutions; and farm organizations for the purposes of:
>
> • Evaluating, changing, promoting, and formulating farm and rural policies and programs that help agricultural producers;
> • Studying historical trends, assessing current conditions, and planning for the future;
> • Formulating market strategies, providing more efficient production and distribution systems, and locating facilities for agricultural communities;
> • Making energy projections and forecasting needs for agricultural producers and their communities;
> • Developing new and improved methods to increase agricultural production and profitability;
> • Allocating local and national funds for farm programs, e.g. extension service projects, agricultural research, soil conservation programs, and land-grant colleges and universities;
> • Planning for operations during drought and emergency outbreaks of diseases or infestations of pests. (USDA – National Agricultural Statistics Service – Census of Agriculture, n.d.)

The reliance upon this data of United States government agencies as well as researchers and non-profit organizations that support and advocate for farmers, especially small farmers is crucial for outreach purposes and to rectify past injustices. It is important especially that there is an understanding of what constitutes a farmer today is one that aligns with our understanding of who is part of the production and supply of our current food system and this idea is aligned with a pursuit of food justice and food sovereignty.

References

2007 Census of Agriculture—History (Volume 2, Subject Series Part 7). (2007). United States Department of Agriculture/National Agriculture Statistics Service. https://www.census.gov/history/pdf/2007aghistory.pdf

2019 County Committee Elections—Farm Services Agency. (2019). *About the U.S. Department of Agriculture | USDA.* (n.d.). Retrieved February 1, 2019, from https://www.usda.gov/our-agency/about-usda

Bigelow, D., Borchers, A., and Hubbs, T. (2016). *U.S. Farmland Ownership, Tenure, and Transfer.* USDA Economic Research Service.

Cummings, J. D. (2016). *Emancipation Proclamation.* Minneapolis, MN: ABDO Publishing.

Dixon, D. & Hapke, H. (2003). Cultivating Discourse: The Social Construction of Agricultural Legislation. *Annals of the Association of American Geographers.* 93(1), 142–164.

Dunbar-Ortiz, R. (2014). *An Indigenous Peoples' History of the United States.* Boston, MA: Beacon Press.

Kendi, I. X. (2019). *How to Be an Antiracist.* New York, NY: One World.

Krom, C. (2010). The Real Story of Racism at the USDA. *The Nation.*

Lincoln's Milwaukee Speech | National Agricultural Library. (n.d.). Retrieved July 8, 2019, from https://www.nal.usda.gov/lincolns-milwaukee-speech

Mason, D. (2009). *The End of the American Century.* Lanham, MD: Rowman & Littlefield Publishers, Inc.

Stevens, K. (2007). Gee's Bend. In *Encyclopedia of Alabama.*

The History. (n.d.). Auburn University. Retrieved February 1, 2019, from http://www.auburn.edu.

U.S. Department of Agriculture: Progress toward implementing GAO's civil rights recommendations. (2012). U.S. Govt. Accountability Office.

USDA – National Agricultural Statistics Service—Census of Agriculture. (n.d.). Retrieved February 19, 2019, from https://www.nass.usda.gov/AgCensus/.

USDA History Collection Introduction/Index | Special Collections. (n.d.). Retrieved February 1, 2019, from https://specialcollections.nal.usda.gov/usda-history-collection-introductionindex.

Vilsack, T. (2010). *MY USDA: A Progress Report for Employees on USDA's Cultural Transformation—Summary of Progress November 2010 through February 2011.* USDA Office of the Secretary.

Washington, Dubois and Johson. (1901). *Three Negro Classics: Up From Slavery (Booker T. Washington); The Souls of Black Folk (W.E.B. DuBois) and The Autobiography of an Ex-Colored Man.* Avon Books.

3 The challenging structure of the U.S.D.A.

As a child, I did not think too much about food in terms of who grew it, harvested it, prepared it for transport, transported it or even what happened to it once it arrived in a grocery store to be sold. My parents belonged to a grocery cooperative in El Cerrito, California, part of the Berkeley Cooperative Collective (Fullerton, 1992) and there was an abundance of fresh produce, but I probably assumed that it all was locally grown. My limited concept of a cooperative at the time, the early 1970s simply meant that we were part of a group of people who, together, owned that grocery store, which also had a credit union. It was quite a neighborhood gathering place and I always looked forward to going shopping there with my mother. According to my mother, there was also a play area for children while whoever brought them there was shopping and apparently there was also a small clothing store where she brought a lot of my clothes when I was in pre-school.

The El Cerrito cooperative grocery store was apparently a very progressive place, connected to a group of cooperative grocery stores. The Berkeley Cooperative will be discussed later in the book because it provides an interesting historical urban example on the concept of cooperatives that some communities and non-profit organizations pursue today in an effort to address food deserts. But an important point is that there were various radical political initiatives that the Berkeley Cooperatives included in their store policies which made it a prophetic feature of the impending food justice movement. The atmosphere at the grocery store had the feel of an extended neighborhood family of people who cared about one another and they even had some type of rewards system for frequenting the store that allowed for occasional discounts on food that was actually reasonably priced. During this same time of the late 1960s – early 1970s the Black Panther Party in West Oakland had their Breakfast Program for Children taking place and a number of Rural farming cooperatives in the South were started, some that have remained active to this date. But this was an urban grocery store, providing healthy food to a predominantly Black neighborhood in northern California designed in a similar spirit of mutual collective responsibility and support and historically situated in the midst of all of the movement activity taking place across the country – Civil, Feminist, Black Power, LGBTQ and Anti-war fervor. Food was central to local community and economic development, regardless of the community.

My perspective on food systems has evolved as I have come to realize the unjust and racialized aspect where there are many people who do not have regular access to healthy foods because they live in what we have come to understand as a "food desert." The U.S.D.A. makes determinations about how food deserts are defined in both urban and rural settings with a measurement of whether a given population in a census tract lives more than one mile away from a grocery store (urban) or more than ten miles from one (rural). The U.S.D.A. also looks at the economic status of the respective census tracts and if it is determined to be low income based on the median family income, to qualify as a food desert tract, at least 33 percent of the tract's population or a minimum of 500 people must have low access to a supermarket or grocery store. (*USDA Defines Food Deserts|American Nutrition Association*, n.d.)

So many situations exacerbate the disparities in what should be a basic right – food – and in what is often contained under vulnerable circumstances, and yet there are many people who have been at the forefront of a food justice movement and are dedicated to making real sustainable changes. People often state that your zip code where you are born and/or raised should not be the determinant of your future success and the same should apply to your access to healthy food. How communities are designed with varying housing options, the availability of public transportation, health care access and good schools are decisions that are not always democratically prescribed. And along with this, there have been so many shifts in the agricultural arena just in the last decade – struggles and also opportunities that draw us together in multiple ways, from both the consumer and producer standpoint.

When two grocery store chains, Marsh and Double 8, abruptly shut down and closed all of their stores in the city of Indianapolis, Indiana there were many neighborhoods in the county, Marion, including my own, that were suddenly thrust into a food desert. The closing of both of these grocery chains had a truly harsh feel about it because it happened at the same time and also because most of the Double 8 food stores were located in lower-income and predominantly Black neighborhoods. Some of the neighborhoods that were directly affected gathered together to figure out what needed to be done differently to come up with more sustainable ways to make healthy food accessible and affordable. Efforts that draw the producer of food and the consumer closer together came to the forefront – leading to a blurring of the rural-urban and suburban lines and a deeper communal focus steeped in food justice and food sovereignty.

Personally, I was able to drive to a number of farmers markets, at least three or four within 15 to 20 miles and there were some grocery stores that were closer, but they still required access to personal transportation, since public transportation does not run in my housing area. The question arises though, what if I did not have my own transportation available? Just the abrupt closing of a grocery store would have been extremely difficult to overcome. This is a question that has to be posed from a collective standpoint, not that of the individual. There are other possibilities, such as grocery delivery services but they can be costly and with services where you can purchase ready-made healthy meals or many other similarly configured food items, there is a steep price, which is hard to maintain for many families. There is always a price for convenience.

The focus on locally grown healthy and accessible food has remained important though some speak as if this is a recent phenomenon, which is not the case. The role of the U.S.D.A. in making sure that our food system, among other things, has an extensive national as well as local reach involves the work of the federal agency in partnership with county offices. There must be accountability between them, however, with an agency as large as the U.S.D.A. it can be difficult. The U.S.D.A. is made up of 29 agencies. Figure 3.1 presents and organizational chart.

The manner in which the U.S.D.A. conducts its business is organized on the local level with the appearance of a democratic structure in the form of county committees that are elected bodies; however, due to the constructed nature of participation, the interests of minority and women farmers have not historically been fully represented. This is where the issue of land ownership versus tenancy and how one is defined as a principal operator of a farm seeps into the political decision-making process that affects all farmers. Because there was little accountability between the county committees, local U.S.D.A. offices and the federal agency, some local practices gave way to the allowance of discriminatory procedures that served to harm many farmers from underrepresented groups, also known as those of a socially-disadvantaged group (S.D.A.) for decades. If you were a minority and/or women farmer in a county that was predominantly white-male farmer dominated it was difficult to win an election and you were furthermore powerless when

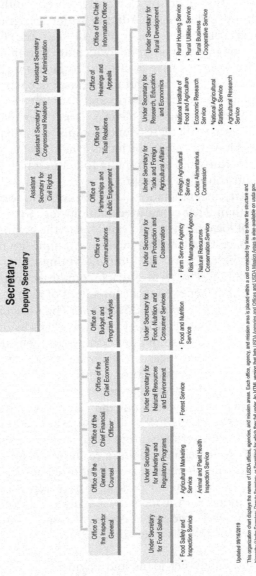

Figure 3.1 U.S.D.A. people's garden sign.

it came to accessing U.S.D.A. farming support. These problems largely took place through the county FSA offices and were at the heart of the allegations of the Pigford, Keepseagle, Garcia and Love settlements.

In the 1930s, County Committees were authorized as part of Agricultural Adjustment Administration programs in relation to the Farm Services Agency (F.S.A.) and the Commodity Credit Corporation (C.C.C.) programs which are administered by the F.S.A. These two agencies have continued to be authorized through Farm Bills, along with other agricultural-related departments and programs such as The Consolidated Farm and Rural Development Act and the C.C.C. Charter Acts (*About the U.S. Department of Agriculture|USDA*, n.d.). The purpose of these county committees is to provide input at the grassroots level involving the local farmers and ranchers to state-level F.S.A. programs and additionally to act in advisory board regarding the administration of these programs.

There are currently more than 7700 committee members who serve on more than 2200 committees nationwide. Committee membership consists of local farmers and ranchers in the community and includes a Chair, Vice-Chair, voting members, a minority non-voting advisor, and in some cases an "appointed," voting minority member designated as S.DA. The term of each committee member is three years and they can be renewed twice for a limit of nine years (*About the U.S. Department of Agriculture | USDA*, n.d.) The Agricultural Credit Act of 1987 (P.L. 100-233) defined S.D.A. individuals as those who may have been subject to discrimination because of their identity as members of a group, without regard to their individual qualities. In addition to racial and ethnic minorities, women are also considered an S.D.A. group. Initially, the targeting was applied only to long-term real estate, or farm ownership, loans, but the Food Agriculture, Conservation, and Trade Act of 1990 (P.L. 101-624) expanded targeting to include operating loans to truly capture those have been affected historically.

The S.D.A. focus followed as a result of the alleged discrimination settlements and there have been more efforts made to ensure that minority and women farmers also have representation at the federal level along with a focus on having equitable access to support at the local level. The stated oversight responsibilities of the County Committee include the following:

1 Income safety-net loans and payments including setting county averages and yields for commodities;
2 Conservation programs;

3 Incentive, indemnity, and disaster payments for some commodities;
4 Energy programs; and
5 Payment eligibility for loans.

The County Committee has a relationship to the state-level committee, in that they conduct hearings and review requests by them and they ensure S.D.A. farmers and ranchers are fairly represented by making recommendations to the state committee with regard to existing programs. They also help to monitor changes to farm programs and provisions coming from the F.S.A. (*About the U.S. Department of Agriculture | USDA*, n.d.).

As a result of the Pigford, Keepseagle, Garcia and Love settlements, there were some access adjustments made to the construction of the F.S.A. County Committees that at least provided that minority and women farmers could not be completely dismissed or marginalized but there were also some loan forgiveness and different procedures put in place considering how different these farmers had been treated for decades. What surfaced during the course of the settlement process was that racial and ethnic minority farmers were forced to use more direct loan programs that guaranteed loan programs due to systemic discrimination. The following was indicated leading up to the settlements:

About 1200 borrowers, or only 3 percent of all borrowers with FSA guaranteed loans, are racial or ethnic minorities. While minorities represented less than 4 percent of U.S. farmers in 1997, they comprised nearly 6,600, or almost 7 percent of all FSA direct borrowers in 1999 excluding lending in Puerto Rico. In some regions, racial and ethnic minorities rely heavily on FSA as a source of capital. In nearly 90 counties where Black farmers are concentrated – in the Mississippi Delta, Southern Coastal Plain from Virginia to Georgia, and arts of the Piedmont – over 25 percent of all Black farmers identified by the 1997 Census of Agriculture had received FSA direct loans since 1993. For many counties on or near Indian reservations, over 25 percent of American Indian farmers were recent FSA borrowers. Likewise, for some counties in West Texas and the Southwest, over 25 percent of Hispanic farmers had obtained an FSA loan in the last 7 years. Because racial and ethnic minorities are more likely to have low average incomes and a limited asset base, they are less likely than other farmers to qualify for credit from private lenders.

(Koenig & Dodson, 1999)

Targeting of loans is accomplished by setting aside a share of the annual loan funding for use by S.D.A. applicants, based on the proportion of S.D.A. farmers or residents in the county or state. Both direct and guaranteed loan programs have targeting requirements. Direct loans are made through F.S.A.'s county and state offices, and F.S.A.-guaranteed loans are originated, funded, and serviced by private-sector lenders. Through both direct and guaranteed loan programs, $296 million was lent to S.D.A. groups in fiscal 1999, about 8 percent of total F.S.A. loan obligations. (Koenig & Dodson, 1999)

Racial and ethnic minority farmers who are at the heart of this book are regionally clustered as explained previously. Hispanic or Latino farmers are to be located in the Southwest, American Indians in the Plains, and Black farmers along the Southern Coastal Plain, parts of the Piedmont and the Mississippi River Delta. Racial and ethnic minority farmers tend to operate smaller operations than non-minority farmers. Only about a third of minority farms reported sales greater than $10,000 in 1997, compared with half for all farms. However, some minority-operated farms are large, mostly Asian-American, bringing the average size to just under $103,000, the average for all farms. Farms operated by Blacks, however, had average sales of $26,000 (*USDA - National Agricultural Statistics Service - Census of Agriculture*, n.d.).

These figures that provided an economic context for the time period covered by the settlement demonstrated the level of marginalization that had occurred over decades on these farming groups. Their land and earning capacity in the agricultural arena have been significantly diminished and yet it is hopeful that all the changes taking place at both the federal and county levels will allow for a certain level of restoration and recovery amongst all of the respective minority and women farmers. Drawing upon the most recent 2017 Census of Agriculture Statistics, a larger picture of the current agricultural arena shows that there are major differences between large and small farms as far as both land mass and earning power as indicated by income. Many people will not be surprised to see that the average age of farmers in the United States is relatively high and that based on the indicated data, there is a shift in how agriculture is being assessed.

Here are some key statistics from the 2017 Census of Agriculture:

• There are 2.04 million farms and ranches (down 3.2 percent from 2012) with an average size of 441 acres (up 1.6 percent) on 900 million acres (down 1.6 percent).

- The 273,000 smallest (1–9 acres) farms make up 0.1 percent of all farmland while the 85,127 largest (2,000 or more acres) farms make up 58 percent of farmland.
- Average farm income is $43,053. A total of 43.6 percent of farms had positive net cash farm income in 2017.
- Ninety-six percent of farms and ranches are family owned.
- Farms with Internet access rose from 69.6 percent in 2012 to 75.4 percent in 2017.
- A total of 133,176 farms and ranches use renewable energy-producing systems, more than double the 57,299 in 2012.
- In 2017, 130,056 farms sold directly to consumers, with sales of $2.8 billion.
- Sales to retail outlets, institutions and food hubs by 28,958 operations are valued at $9 billion.
- The average age of all producers is 57.5, up 1.2 years from 2012.
- There are 321,261 young producers age 35 or less on 240,141 farms.
- One in four producers is a beginning farmer with 10 or fewer years of experience and an average age of 46.3.
- Thirty-six percent of all producers are female and 56 percent of all farms have at least one female decision maker. Farms with female producers making decisions tend to be smaller than average in both acres and value of production.
- Female producers are most heavily engaged in the day-to-day decisions along with record keeping and financial management (*USDA - National Agricultural Statistics Service - Census of Agriculture*, n.d.).

One key difference in information provided by the most recent U.S. Census data is with regard to how women farmers are counted or remarked upon in the data. Previously they were typically just seen as appendages to their husband's farm where the previous counting policy indicated that only one person could be counted as a principal operator. This was another change that came out of the settlements to make it easier for women farmers, in particular, to access resources and support directly from the U.S.D.A. With the re-establishment of the Civil Rights Division within the U.S.D.A. at the federal level and more awareness and attention given to the decisions of the County Committees, there have been a lot of changes in the ability of minority and women farmers to access support for their farming efforts, which has allowed for some creative approaches, many steeped in cultural traditions, to be used for young farmers of color and young women farmers to become viable participants in the agricultural arena. These efforts will be highlighted later on in the book.

References

About the U.S. Department of Agriculture | USDA. (n.d.). Retrieved February 1, 2019, from https://www.usda.gov/our-agency/about-usda.

Fullerton, M., (ed.) (1992). *What ever happened to Berkeley Co-Op: A collection of essays.* Madison, WI: The Center for Cooperatives.

Koenig, S. and Dodson, C. (1999). *FSA Credit Programs Target Minority Farmers* (Agricultural Outlook).

USDA - National Agricultural Statistics Service—Census of Agriculture. (n.d.). Retrieved February 19, 2019, from https://www.nass.usda.gov/AgCensus/.

USDA Defines Food Deserts | American Nutrition Association. (n.d.). Retrieved February 11, 2019, from http://americannutritionassociation.org/newsletter/usda-defines-food-deserts.

4 Pigfords I and II – The case involving Black farmers

This chapter discusses the Pigford settlement with the United States Department of Agriculture (U.S.D.A.) which involved Black farmers and will show that blatant discrimination as well as other factors contributed to their significant land loss leading to a substantial decrease in the number of Black farmers today. Discussed in both historical and contemporary contexts will include how Black farmers developed strategies such as cooperatives and selling their products in town squares, similar to today's farmer's markets to negate their demise but also they used their land holdings as a foundation for addressing food and social injustices. A lot of organizing for the Civil Rights and Voting Rights Movement in the rural South took place on the private land-holdings of Black farmers. One example is that there are markers along the Selma-to-Montgomery, Alabama Voting Rights trail with these indicators (*Alabama: Selma to Montgomery National Historic Trail*, n.d.). Young farmers today, from diverse racial and economic backgrounds use similar self-preservation strategies as part of their social justice efforts towards making healthy food more accessible and as a means toward sustainable community economic development.

This is not just the story of land loss suffered by Black people in the United States, especially Black farmers, though that is an important story itself, but we must also consider what type of adjustments were made towards food justice efforts conceived through adversity. Additionally, the struggles that have helped shaped current innovative ways of farming today do have roots that are tied to other movements, such as the Civil Rights Movement and Voting Rights Movement, the labor movement and movements related to environmental justice. There has been a collective will of Black farmers who have been motivated to persevere as long as possible regardless of the lack of governmental support and even through the usual difficulties that farmers face. They were at the forefront of advocating for redress from all of the discrimination they suffered that ultimately led to the Pigford settlement which subsequently opened the door for the other settlements, like Keepseagle to follow. But it is important to continue to place their struggle into the historical context beginning with the institution of slavery and the agricultural skills set the enslaved Africans brought with them.

In her book, "Freedom Farmers: Agricultural Resistance and The Black Freedom Movement," Monica White offers, "The millions of men, women and children who were kidnapped from their homelands in Africa and transported through the Middle Passage, in the most extreme case of forced deportation in world history, possessed knowledge of microclimates and the particular kinds of crops that they could grow in the places where they would toil as enslaved" (White, 2018). The recognition of the skills they brought with them is critical to understanding this intergenerational transfer of knowledge that has been deliberately thwarted at various time periods in American history.

> "Slavery had a tremendous influence on food systems around the world. Enslaved Africans were highly skilled farmers who not only grew rice, cotton, sugar and tobacco, but were also expected to grow food for themselves as well as the plantation owners, for whom they also had to cook. The famed southern cooking and "soul food" of the United States is an African American invention with deep roots in slavery."
>
> (Holt-Gimenez, 2017)

W.E.B. DuBois estimated 19th century progress in land ownership by Black farmers: 3 million acres in 1875, 8 million in 1890 and 12 million in 1900 (Washington, Dubois, & Johson, 1901). The Census of Agriculture shows a steady increase in the number of farm operators owning land in the South from 1880 to 1890 and again in 1900 but does not distinguish between white and nonwhite owners until 1900. Census figures show 1920 as the peak year in the number of nonwhite owners of farmland in the South. In terms of acreage owned, the census shows 1910 as the peak year for the South. More than 12.8 million acres were fully and partly owned, respectively, by 175,290 and 43,177 nonwhite farmers. The peak for the existence of Black farmers was 1920, according to the U.S.D.A., at 925,710 (*USDA History Collection Introduction/Index | Special Collections*, n.d.).

Due to the challenges faced historically by Black farmers to receive the same type of resources as white farmers, from local agencies of the

U.S.D.A., they needed to come up with alternative structures in order to survive. So they created farming collectives and cooperatives, came up with ways to directly sell to the public, such as setting up in town squares, developed urban gardens in connection to migration patterns and transplanted farmers, and used organic farming and some soil conservation farming practices. These are ideas and values that we will see across all of the groups that are a focus of this book. "Slavery had a tremendous influence on food systems around the world. Enslaved Africans were highly skilled farmers who not only grew rice, cotton, sugar and tobacco, but were also expected to grow food for themselves as well as the plantation owners, for whom they also had to cook. The famed southern cooking and "soul food" of the United States is an African-American invention with deep roots in slavery" (Holt-Gimenez, 2017).

John Hope Franklin wrote,

"The last quarter of the nineteenth century witnessed the steady deterioration of the position of Negroes in the United States. The end of Reconstruction had left them without any protection from the merciless attacks of the Klan and other terrorist groups who continued to use the mythical threat of "Negro rule" as their excuse for lawlessness. ...In the South the new industries were closed to them; and most were forced to subsist as small farmers or hapless sharecroppers. In the North new labor unions barred them from membership, while "new" immigrants showed their hostility in a variety of ways."

(Washington et al., 1901)

Increases in land ownership after 1900 were partly due to a significant rise in cotton prices that lasted until the outbreak of World War I in 1914. The growth in farmland acquisition by Blacks during the late 19th and early 20th centuries demonstrates a period of economic mobility for about 25 percent of farm operators (Washington et al., 1901). Established in 1865 by the United States Congress, the Bureau of Refugees, Freedman and Abandoned Lands, also known as "The Freedman's Bureau" was charged with helping millions of enslaved Black people and poor whites after the ending of the Civil War. Unfortunately, the Freedman's Bureau was prevented from fully carrying out its goals but it initially attempted to set up schools and provide some medical aid and create tenancy arrangements under rental contracts and sharecropping that were supposed to ultimately lead to Black people becoming self-sufficient. For example, during the reconstruction period and subsequent decades there was a substantial growth in farm-operating arrangements but that worsened with the rise of the Jim Crow era in the 1890s. The well-known Special Field Order, No. 5 issued by Union General William Tecumseh Sherman, a temporary plan that granted each freed family 40 acres of tillable land on islands and the coast of Georgia, along with a number of unneeded army mules, was only in effect for one year and, therefore, not truly enforced. The tenant contracts were often not a fixed-rent type, but a specific share of either the harvest of sales. In contrast to sharecroppers, tenants supplied more farm production inputs in addition to their labor. The distinction had originally meant that tenants paid landowners for use of the land, including debt payments, while sharecroppers received their share, less debt payments, from landowners (Stack, 1996).

During the 19th century, there were some opportunities to establish farms on unsettled lands, but over the long run, most Black farmers gained land through their working relationship with white planters (Finkelman, 2009). Landowners profited by offering tenant farm operators the incentive of having an opportunity to buy certain tracts of land in exchange for increased farming efficiency. And yet a lot of these contractual agreements were problematic as they often took advantage of the limited literacy level of Black people at the time and thus they were trapped from one generation to the next in debt that was impossible to overcome. Enactment of Jim Crow laws in the late 1890s empowered landlords and planters to try to extract more output from tenants and sharecroppers with less compensation, rather than using incentives for self-motivated work. Oppressive farm operating contracts were easier to impose because the voting rights of Black people were limited by law and by violence. Without the franchise, black tenants and sharecroppers had no legal or political recourse. These laws also facilitated tacit coordination by white landlords in applying stricter terms in agricultural contracts (Washington et al., 1901).

Seeking support for farming operations from the Farm Services Agency (FSA), many Black farmers had their applications for loans torn up right in front of them, disregard and refusal of support, denial of bank loans, underfunding and as a result, suffered traumatic loss of their land and livelihood. Black farmers reported that local loan authorities in all-white county committees in the South spitting on them, throwing their loan applications in the trash and illegally denying them loans, which happened blatantly for decades up through the 1990s. When their loans were approved, the process was slow; they were subjected to line-by-loan micro-managed approval process as opposed to white farmers who were given unsupervised on-time loans (Krom, 2010). It was after being subjected to this discrimination for decades that Black farmers came together collectively, created organizations like the National Black Farmers Association and decided to sue the government (Boyd, 1995).

The discrimination case against the U.S.D.A. involving Black farmers was named after Timothy Pigford, who filed a class-action lawsuit in the United States District Court for the District of Columbia on August 28, 1997, against then Secretary of Agriculture Dan Glickman, alleging that the U.S.D.A. discriminated against African-American farmers by denying or delaying applications for benefit programs and loans and that there was also a mishandling of discrimination complaints. Court documents state that 401 African-American farmers from Alabama, Arkansas, California, Florida, Georgia, Illinois, Kansas, Missouri, Mississippi, North Carolina, Oklahoma, South Carolina, Tennessee, Texas and Virginia allege, "that the U.S.D.A. willfully discriminated against them when they applied for various farm programs, and that when they filed complaints of discrimination with the U.S.D.A., the U.S.D.A. failed to properly investigate those complaints" (Cowan & Feder, 2010).

Even further court documents state, "Plaintiffs challenge the U.S.D.A.'s administration of several different farm loan and subsidy programs and/or agencies. Until 1994, the U.S.D.A. operated two separate programs that provided price support loans, disaster payments, "farm ownership" loans and operating loans: the Agricultural Stabilization and Conservation Service (ASCS) and the Farmers Home Administration (FmHA). In 1994, the functions of the ASCS and the FmHA were consolidated into one newly-created entity, the Farm Service Agency (Cowan & Feder, 2010).

Black farmers who farmed between January 1, 1983, and February 21, 1997, and applied for participation in federal farm programs with the U.S.D.A. at that time and believed they were discriminated against on the basis of race by the U.S.D.A. filed a written discrimination complaint with the U.S.D.A. in that time period. The discrimination was institutionalized because of the process required by farmers seeking a loan or subsidy from the FSA had little oversight and/or accountability as to how they were making the decisions and their unfair practices. The applications of these Black farmers would be submitted to a county committee that was comprised of farmers from that respective county who are elected by other farmers in that county. If the county committee approved the application the farmer would receive the subsidy or loan, however if the application was denied the farmer was allowed to appeal to a state committee and then to a federal review board and yet it was the make-up at all of those levels which was problematic as they allowed for little participation of Black farmers or other underrepresented groups (Cowan & Feder, 2010).

There was also a County Executive Director that was supposed to work with farmers on their applications. There were few, if any African-Americans that were ever elected to serve on county committees, which served as a mechanism to further entrench discriminatory practices and enshrine a good-ole-boy's network that to this day is still hard to change. Through pressure from the U.S.D.A., some counties actually would appoint a token ex-officio Black farmer to the committee but that individual was not given the power to cast any votes so their presence was simply ineffective. There is still a non-voting SDA position on the County Committees, however under certain circumstances, especially if an FSA office has had track record of disparities (*2019 County Committee Elections – Farm Services Agency*, 2019).

Three types of projects, during the period 1880–1932 were crucial: cooperatives, farm settlement projects and farming self-sufficiency. Then there was the New Deal period 1933–1941, where the government was instrumental in turning over some land to African-American communities, such as in the Gees Bend, Alabama area, to secure their land ownership and economic development, while at the same time this served as the heyday of the rise of the Ku Klux Klan. There was then the organizing and activism of the Civil Rights Movement and the rise of Black farmer cooperatives during the 1950–1960s with the crucial role of the Federation of Southern Cooperatives/Land Assistance Fund, which then has been subsequently served by the rising influence of conservative politics that focused on, among other things, reversing some of the civil rights legislation that shaped black progress and economic opportunities and allowed for crucial governmental civil rights agencies to be eliminated.

Looking back, in 1983, under the directives of then President Ronald Reagan, the U.S. Department of Agriculture Office of Civil Rights Enforcement and Adjudication (O.C.R.E.A.) was abolished. In testimony before a subcommittee actually held at the National Underground Railroad Freedom Center, John W. Boyd, Jr. President of the National Black Farmers Association stated, "...African-American and other socially disadvantaged farmers were left with little hope of any resolutions for civil rights complaints. At the time there were two people working on cases of employment discrimination and not one U.S.D.A. (United States Department of Agriculture) staffer assigned to work on black farmer discrimination complaints and cases. Piles and boxes of complaints, literally with years of dust growing on top of them, went virtually unprocessed, uninvestigated" (Chabot, 2005).

Jurisdiction for the Civil Rights Division covered the FSA – the primary department responsible for discriminatory practices, and therefore, once eliminated, the ability to seek redress was severely crippled, along with subsequently the ability to develop and/or sustain fully operable and competitive farms. And further, because there existed system-wide marginalizing, discriminatory and oppressive conditions at the local level through these county-based departments of the FSA, the level of civil rights violations continued to increase and become somewhat institutionalized as the normal par for course.

A time frame for these Black farmers seeking some type of resolution from their complaints did not in any way coincide with their seasonal needs to plant and harvest, which naturally affected their ability to be successful, carn any profit and even hold on to their land. In court documents it is also stated, and was further verified that there was "a complete failure by the USDA to process discrimination complaints and once OCREA was dismantled complaints were never processed, investigated or forwarded to the appropriate agencies for conciliation. As a result, farmers who filed complaints of discrimination never received a response or if they did receive a response it was a cursory denial of relief."

(Cowan & Feder, 2006)

So finally, the Pigford settlement was agreed upon in 1998 and with it a two-track process, known as Tracks A and B. This set the tone and stakes for the Keepseagle, Garcia and Love settlements. Track A was an automatic payment of $50,000 designed to provide an expeditious resolution to farmers accepted as class members who had no documented evidence to substantiate their claims. There was a facilitator in place to process these claims and it was supposed to result in a maximum time of submission of claim to the decision of 110 days (Cowan & Feder, 2006).

The larger issues surrounding this case in the historical context as described previously concern issues of land ownership and relative power, collective community sustainability and at the very basic level food security. In remarks made at a Black Farmer's Demonstration in March of 1999 in Washington, D.C., Congresswoman Eva M. Clayton stated the following:

> "There is reason to despair because, in my home state of North Carolina, much like every state where farming is a way of life, there has been a 64% decline in minority farmers, in just over 15 years from 6,996 farms in 1978 to 2498 farms in 1992. Black farmers are declining at three times the rate of white farmers. There are several reasons why the number of black farmers is declining so rapidly. But, the one that has been documented, time and time again, is the discriminatory environment present in the Department of Agriculture, the very agency established to accommodate and assist the special needs of farmers."
>
> ("Written Testimony of John W. Boyd, Jr. President, National Black Farmers Association." – "Civil Rights in Light of Pigford v. Glickman," 2005)

African-American farmers make up just a little over one percent of all farmers and yet they do n't even receive one percent of all crop subsidy payments. Looking back at a closer time period related to the Pigford Settlement, the 2007 Census of Agriculture on Black Farmers found that the majority had an average size farm of 104 acres, compared to the national farm average of 418 acres and their average sales are $21,340 compared to the national average for all farms of $134,807. The report also states that "Almost half (46 percent) of all black-operated farms are classified as beef cattle farms and ranches, compared to 30 percent of all farms nationwide." Black farmers also tend to be older with an average of 60.3 years, as compared to 57.1 years for U.S. farmers overall and a total of 37 percent of all Black farmers are 65 or older compared to 30 percent of all farmers nationwide (*USDA – National Agricultural Statistics Service – Census of Agriculture*, n.d.). There is also a crucial disparity in terms of the percentage of black farms that actually have internet access, important in having the ability to seek out additional opportunities and information that provide support to their farming endeavors. In an interview in the "New Journal and Guide," John Boyd, Jr., President of the National Black Farmers Association (NBFA) points out that many Black farmers typically face three particular systematic problems: 1) access to credit, 2) access to good seed and 3) the ability to get good competitive prices for their crops (*"Written Testimony of John W. Boyd, Jr. President,*

National Black Farmers Association." - "Civil Rights in Light of Pigford v. Glickman," 2005).

Since the initial Pigford settlement there have arisen many additional issues as a result of the adjudication process, which have been characterized as willful obstruction and undermining of the justice that was overdue for these farmers at the hands of the U.S.D.A. itself. In 2002, the U.S.D.A. Secretary Ann M. Veneman developed an action plan to strengthen programs aimed at serving minority and disadvantaged farmers as part of an ongoing effort to address concerns of Black farmers. She stated, "We believe these actions will provide additional focus on our efforts to ensure fair and equitable treatment for all producers." These measures, along with a strong work plan for implementation, incorporate many of the recommendations made by Black farmers in recent months" (Veneman).

Some of the directives the action plan outlined were the following:

1 An internal working group to ensure effective coordination of resources for minority farmers across various U.S.D.A. agencies...
2 No acceleration of loan repayment or foreclosures on borrowers who have claims pending under the Pigford Consent Decree.
3 Transferring of $100 million in additional funds to FSA's direct operating loan program to assist minority, small, beginning, limited resource and other farmers.
4 Customer service training for FSA state and local managers and employees to emphasize the importance of a more timely loan processing.
5 Establishment of the Office of Minority and Socially Disadvantaged Farmer Assistance.
6 And, diversity training for employees who have direct contact with constituents throughout the country, particularly in the local offices...among other changes.

A report by the Environmental Working Group at the time, funded by the Ford Foundation discovered that 86 percent of African-American farmers who came forward seeking restitution initially were denied – 81,000 out of 94,000 farmers. The report indicated that there were over 22,000 farmers accepted into the Track A and just a little over 200 under Track B. And yet there were still close to about 74,000 that sought entry through a late-claims process having missed the initial deadlines established. There was admitted problems with notification of the class action suit including the process for seeking restitution.

In their report, the Environmental Working Group found that ultimately "of the nearly 100,000 farmers who came forward with racial discrimination complaints, 9 out of 10 were denied any recovery from the settlement. As a result, instead of the $2.3 billion, U.S.D.A. only provided $650 million in direct payments to farmers." There was a little over 200 farmers who opted out of the settlement and had better results in prevailing (*Obstruction of Justice: USDA Settlement Fails Black Farmers*, 2004).

Another important aspect of the problems surrounding the initial Pigford settlement was that the U.S.D.A. had contracted with the United States Department of Justice (USDOJ) to provide legal representation which resulted in, "56,000 hours of attorney and paralegal time challenging 129 farmers' claims...This amounts to an average of 460 hours, or nearly 3 months of time, devoted to contesting each individual farmers' claim. Assuming an average salary of $60,000... this represented a potential cost of $330 million to taxpayers, the equivalent to the cost of providing Track A relief for 6,600 farmers" (*Obstruction of Justice: USDA Settlement Fails Black Farmers*, 2004). Because so many farmers relied upon securing documents that they had submitted to the U.S.D.A. to prove their claims of discrimination, because they had to prove that a similarly situated white farmer was provided loans and subsidies that they were denied, a system of obstruction ensued in order to prevent these farmers from actually obtaining these documents. The U.S.D.A. blatantly denied requests from these farmers that were submitted under the Freedom of Information Act.

Seeking redress for the thousands of farmers, who were even denied the opportunity to seek justice in the initial settlement class action process, then became the basis for Pigford II, an attempt to ensure that all eligible Black farmers were heard. Pigford II was allowed through a provision in the 2008 farm bill that permitted any farmer who had submitted a late-filing under Pigford but had not received a determination on the merits of their claim to petition in federal court for such a determination. As pointed out in a Congressional Research Paper, "A maximum of $100 million in mandatory spending was made available for payment of these claims." In February of 2010, Attorney General Holder and Secretary of Agriculture Vilsack (the third Agriculture Secretary with regard to the plight of the Black farmers) announced a $1.25 billion settlement of these Pigford II claims, contingent upon congressional approval. It was not until the Senate passed the Claims Resolution Act of 2010 (H.R. 4783) by unanimous consent with the House also passing this Senate bill then it was signed into law in December by President Barack Obama. In addition, under subsequent new leadership, the U.S.D.A. instituted a number of policy changes at the national level to provide more systematic support to, not just Black farmers, but to all minority farmers and women farmers. More difficult, however, has been the ability to change the accessibility of loans and subsidies at the front-line level due to the locally-elected and controlled county commission still in place that controls this process.

A follow-up report by the GAO in 1999 found that 44 percent of program discrimination cases and 64 percent of employment discrimination cases had been back logged for over a year, "Many farmers who joined the lawsuit were also denied payment. By one estimate nine-out-of ten farmers who sought restitution under Pigford were denied. The Bush Department of Justice spent 56,000 office hours and $12 million contesting farmers' claims. Many farmers felt their cases were dismissed on technicalities." (Krom, 2010).

The most recent 2017 Census of Agriculture data indicates that the number of Black farmers is up about 2 percent from the 2012 Census of Agriculture data at 45,508, 95 percent of all farmers are white at 3.2 million (*USDA – National Agricultural Statistics Service – Census of Agriculture*, n.d.). There are striking differences in the amount of acreage per farm between white and Black farmers as well as earned income and profits. However, the slight increase in Black farmers from the 2012 Ag Census count to the most recent 2017 Ag Census suggests that some of the initiatives that came out of the Pigford settlement with regard to outreach efforts from the U.S.D.A. along with an enhanced focus on the significant work of Black and other marginalized farmers are starting to take some positive effect, if not serve as stop-gap measures. Another helpful shift is that the Census of Agriculture now defines farmers in terms of producers and operators, which broadens the perspective of those who are participating in varying forms of agriculture.

Looking at current income levels only 2349 Black farmers in 2017 were running farm operations that made more than $50,000 a year, compared with 492,000 white farmers. Internet connectivity, which is critical for marketing these days, was indicated at 61 percent for Black farmers in the 2017 Ag Census, compared with 76 percent for white farmers. So many young farmers have the advantage of technology and expertise in social media that helps them survive the instability of small farm operations, though there are fewer younger/newer farmers in comparison. In fact, the average age of farmers has increased since the last Census of Agriculture (*USDA – National Agricultural Statistics Service – Census of Agriculture*, n.d.).

Throughout their fight for justice, there was the ongoing fight simply to survive that preceded the Pigford lawsuit. Black farmers faced the usual risky challenges associated with farming generally that were tied to the environment but they also faced the challenge of direct violence at the hands of white people who felt that Black people simply should not be land owners. In the Black Belt counties of Alabama which was one location that served as the frontline for civil and voting rights movement activism, once the Voting Rights Act of 1965 was passed, Black people used their collective votes to make real changes in county-level positions. It is at the county level that so much direct decision-making takes place that affects people's lives and among the main positions they sought to secure were Tax Assessor and Tax Collector. It was their hope that a fair system would be put in place to prevent the government from assessing their land at two high of a value so that they couldn't pay taxes and their land would be auctioned off. They also understood the power of the Tax Collector. The other position Black people sought at the County Level was the Sheriff. Considering other well-known traditionally white Alabama Sheriff's like Bull Connor of Birmingham, Alabama or Lummie Jenkins from Camden, Alabama, voting in a young Black Sheriff in the early 1970s at that time, named Prince Arnold allowed for much better relations between law enforcement and the predominantly Black rural community.

Along with these Black farmers who were the focus of the Pigford case, there have been many more contemporary Black farmers, some fairly young and a number of women farmers who have created their own unique ways of farming and of largely participating in the agricultural arena. One fairly prominent example is a young woman by the name of Leah Penniman" who wrote the bestseller "Farming While Black," who I listened to on a Young Farmers podcast discusses her work at "Soul Fire Farm" which she describes as "Black and Brown-centered community farm that is dedicated to ending racism and injustice in the food system." She stated the following:

> Black people have a history in regenerative agriculture that is not circumscribed by slavery, sharecropping and tenant farming. We have a tens of thousands of years old history of innovating and coming up with dignified solutions to solving hunger in our community without destroying the planet.

The main things that Leah and her husband and staff do at Soul Fire Farm are described by her as:

1 *Run a working commercial farm that provides vegetables, eggs, fruits and poultry and center the needs of refugees and immigrants and those impacted by mass incarceration and police violence in their farm share and food distribution network. Use Afro-indigenous practices, heirloom seeds and committed to leaving the earth and the land better than we found it.*

2 *Training, equipping and resourcing the next Black and Brown farmers. Day to full-season training programs – generation reclamation of the land.*

3 *Organizing to change the structures that hold up and bolster this racist food system – work regionally and nationally land reparations and policy shifts and other initiatives to make sure that everybody, regardless of background has access to land and can be a farmer with dignity and consume culturally appropriate and healthy foods.*

In her work, Penniman draws upon some of the same strategies and survival techniques of the Black farmers and additionally frames her work in connection to the agricultural arena as one that is focused on addressing systems of injustice and oppression in very deliberate ways. Additionally, she uses her farm as a training place for others and thus engaging in a collective response to addressing food sovereignty.

Collective responsibility and support in the form of cooperatives is another way that Black farmers have survived through their limited resources and adversity. One such organization is the Federation of Southern Cooperative Land Assistance organization. They have been around for 50 years and, are (democratically controlled and owned business) there are both economic and political rationales. They have established an Emergency Land Fund to help address land loss particularly among Black farmers and they look at their programs to generate youth interest and support for cooperatives that need to consider succession plans.

References

2019 County Committee Elections—Farm Services Agency. (2019).

Alabama: Selma to Montgomery National Historic Trail. (n.d.). Natioal Park Service.

Boyd, J. (1995). History—National Black Farmers Association [Http://www. nationalblackfarmersassociation.org/].

"Written testimony of John W. Boyd, Jr. President, National Black Farmers Association."—"Civil Rights in Light of Pigford v. Glickman", (2005) (testimony of Chabot, Steve).

Cowan, T. & Feder, J. (2006). Pigford case: USDA Settlement of a Discrimination Suit by Black Farmers Note. *Pigford Case: USDA Settlement of a Discrimination Suit by Black Farmers*, 1–6.

Cowan, T. & Feder, J. (2010). *The Pigford Cases: USDA Settlement of Discrimination by Black Farmers [Congressional Research Service]*.

Hinson, W. R. (2018). Land gains, land losses: The Odyssey of African Americans Since reconstruction. *American Journal of Economics and Sociology*, 77(3–4): 893–939. https://doi.org/10.1111/ajes.12233

Holt-Gimenez, E. (2017). *A foodie's guide to capitalism: Understanding the political economy of what we eat*. New York, NY, USA: Monthly Review Press.

Krom, C. (2010). The real story of racism at the USDA. *The Nation*.

Lindsey. (2018). *Interview with Cornelius Blanding of the Federation of Southern Coopertives and Land Assistance Fund*. Young Farmers Podcast.

Obstruction of Justice: USDA Settlement Fails Black Farmers. (2004). Envronmental Working Group (EWG).

Stack, C. (1996). *Call to Home: African Americans Reclaim the Rural South*. Basic Books.

USDA – National Agricultural Statistics Service – Census of Agriculture (n.d.). Retrieved February 19, 2019, from https://www.nass.usda.gov/AgCensus/

USDA History Collection Introduction/Index | Special Collections (n.d.). Retrieved February 1, 2019, from https://specialcollections.nal.usda.gov/ usda-history-collection-introductionindex.

Washington, B. T., Dubois, W. E. B., & Johson, J. W. (1901). *Three Negro Classics: Up From Slavery (Booker T. Washington); The Souls of Black Folk (W.E.B. DuBois) and The Autobiography of an Ex-Colored Man*. Avon Books.

White, M. M. (2018). *Freedom Farmers: Agricultural Resistance and The Black Freedom Movement*. University of North Carolina Press.

5 Situating the Keepseagle settlement in sovereign relations

The film "The West: The Geography of Hope," provides insight on the expansion of the United States territory through western conquests from an understanding of how Indigenous people were harmed, marginalized and disregarded primarily through the homesteading process. Lord James Bryce is quoted as stating "Men seem to live in the future, not in the present. They see the country, ten, 50, 100 years hence." He was referring to more recent European immigrants who saw their own participation in this United States national development process, regardless of the Indigenous nations who already occupied the land. They also seemed to have determined that the future of the country was their power alone to determine. The film shows that by 1877 the American conquest of the west was nearly complete, and between 1877 and 1887, four and a half million people came to the western part of the country to settle, as they suggested, "some seeking freedom while others found a place to start over, change themselves." (Ives, 1996)

Further shown in the film is how the Union Pacific Railway was complicit in luring people to homestead in the west, namely places such as Kansas, Nebraska, Colorado, Wyoming and the Dakotas. While this was difficult terrain, they were able to lure about two million people to settle there, which was a serious disruption to the lives of the Native American nations that already occupied those areas, or were relocated to some of those places. This began an ongoing process and heightened time of systemic violence toward many of the indigenous nations as not only were they moved to confined reservations which many of them referred to as "prisons," due to the disruption of their natural roaming culture over the great plains, but there were also many efforts both public and private to move them toward assimilation into a predominantly white-cultural, Americanized identity. This ultimately involved the establishment of the Bureau of Indian Affairs through the United States Department of Interior, and the establishment of schools for Indian children where they were ordered to give up their first language, speak English and also to give up their faith and adopt a foreign Christian religion.

In the film, they quote Sitting Bull of the Lakota nation stating, "I would rather die an Indian than live as a white man." 1877 is a key date because this was also the year that the last of the federal troops were withdrawn from the south. Moving forward, many Black families decided to head west to escape the immediate institutionalization of Black codes that limited their freedom, along with the unfair practices of the sharecropper and tenant agreements to which they were subjected. In addition, there was the terrorism of the Ku Klux Klan against which they had no governmental protection and no freedom to even protect themselves. So post-Civil War and moving into the early 20th century, we have growth of the United States that was shaped by the massive displacement of Native Americans, the emancipation of fairly recently enslaved Black people and a significant influx of recent European immigrants, all collectively seeking stability and liberation from oppressive conditions.

In her book "The Indigenous People's History of the United States," Roxanne Dunbar-Ortiz details the struggles of many Indigenous nations to hold on to their land against the U.S. government's land expansion ambitions. She writes

"The 371 treaties between Indigenous nations and the United States were all promulgated during the first century of US existence. Congress halted formal treaty making in 1871, attaching a rider to the Indian Appropriation Act of that year stipulating 'that hereafter no Indian nation or tribe within the territory of the United States shall be acknowledged or recognized as an independent nation, tribe or power with whom the United States may contract by treaty." She further rights, "During the period of US-Indigenous treaty-making, approximately two million square miles of land passed from Indigenous nations to the United States, some of it through treaty agreements and some through breach of standing treaties."

Under the "Doctrine of Discovery," the rights of Indigenous nations were negated and they were only allowed the right of sale to "European nations" who claimed to have "discovered" the land. This was endorsed as the "will of God" by Western Europe Christian churches and a process of land acquisition was then pursued under the guise of a "trusteeship," which led to policies that allowed for the United States to hold the land in trust "for' them" with some upwards of two billion acres of land stolen under this process (Dunbar-Ortiz, 2014). This was at the heart of what is known as the "Cobell" settlement focused on Native American land held in trust by the U.S. Department of Interior that, along with Pigford, set the stage for the Keepseagle settlement focused on Native American farmers and their relationship with the U.S. Department of Agriculture. The two separate and unique contemporary cases involving Native American land usage and sovereignty are known as Cobell v. Salazar and Keepseagle v. Vilsack. The former centers on claims by thousands of Native Americans that the federal government mismanaged billions of dollars in oil, gas, grazing, timber and other royalties overseen by the Department of the Interior for Indian trustees since 1887 and is named after Elouise Cobell, a Blackfeet activist. The latter involves thousands of tribal plaintiffs who contend that the Department of Agriculture officials denied or delayed a number of farm and ranch loans and emergency assistance applications by Indians which was more in line with the experiences of the Black farmers.

Cobell had been ongoing since 1996 while Keepseagle since 1999. Many of the Native Americans who would have benefited from the settlement in both cases have passed away, according to their lawyers and plaintiffs but the agreement called for $1.4 billion for individual Indian trust fund beneficiaries and $2 billion for a land consolidation program to be overseen by Interior to buy back fractionated trust lands.

First filed in 1996, Cobell v. Salazar involves the Department of the Interior's (DOI's) management of several money accounts. These money accounts, known as IIMs (an abbreviation for Individual Indian Monies) are monies which the federal government holds for the benefit of individual Indians rather than property held for the benefit of an Indian tribe. The conflict in the case emanates from the federal government's trust responsibility with respect to American Indians. The Supreme Court first formulated the concept of the federal government as trustee for Indian tribes in 1831, likening the relationship to that of "a ward to its guardian." In the capacity of trustee, the United States holds title to much of Indian tribal land and land allotted to individual Indians. Receipts from leases, timber sales, or mineral royalties are paid to the federal government for disbursement to the appropriate Indian property owners. Flowing from the federal trusteeship of Indian property are fiduciary responsibilities on the part of the United States to manage Indian monies and assets which have been derived from these lands and are held in trust.

(Garvey, 2010)

The Cobell settlement was approved by Congress on November 30, 2010 (Claims Resolution Act of 2010) and signed by President Obama on December 8, 2010. The $3.4 billion Cobell Settlement includes a $1.9 billion Trust Land Consolidation Fund and $1.5 billion in direct payments to class members. The Cobell Settlement Agreement resulted in several notable outcomes:

- **The Land Buy-Back Program For Tribal Nations:** The Secretary of the Interior established the Land Buy-Back Program for Tribal Nations to implement the land consolidation provisions of the Cobell Settlement Agreement. The Settlement provided for a $1.9 billion Trust Land Consolidation Fund to consolidate fractional land interests across Indian Country.
- **National Commission on Indian Trust Reform:** As part of President Obama's commitment to fulfilling this nation's trust responsibilities to Native Americans, the Secretary of the Interior named five prominent American Indians to a commission to undertake a forward-looking, comprehensive evaluation of Interior's trust management of nearly $4 billion in Native American trust funds.
- **Individual Settlement Payments:** The Settlement provided for one-time payments to anyone who had an Individual Indian Money (IIM) account anytime from approximately 1985 through September 30, 2009 or had an individual interest in land held in trust or restricted status by the U.S. government as of September 30, 2009.
- **Indian Education Scholarship Fund:** The Settlement created the Indian Education Scholarship Fund to help "defray the cost of attendance at both post-secondary vocational schools and institutions of higher education." (*Cobell v. Salazar*, 2009)

In relationship to the ongoing processes that were occurring in connection to both Cobell and Keepseagle, Congress actually passed and President Barack Obama signed a Native American Apology Resolution, a portion which is offered here with the actual resolution presented in totality due to the last lines of the statement:

Whereas the Federal Government violated many of the treaties ratified by Congress and other diplomatic agreements with Indian tribes;

Whereas the United States forced Indian tribes and their citizens to move away from their traditional homelands and onto

federally established and controlled reservations, in accordance with such Acts as the Act of May 28, 1830 (4 Stat. 411, chapter 148) (commonly known as the "Indian Removal Act");
  Whereas many Native Peoples suffered and perished—

1 during the execution of the official Federal Government policy of forced removal, including the infamous Trail of Tears and Long Walk;
2 during bloody armed confrontations and massacres, such as the Sand Creek Massacre in 1864 and the Wounded Knee Massacre in 1890; and
3 on numerous Indian reservations;

Whereas the Federal Government condemned the traditions, beliefs, and customs of Native Peoples and endeavored to assimilate them by such policies as the redistribution of land under the Act of February 8, 1887 (25 U.S.C. 331; 24 Stat. 388, chapter 119) (commonly known as the "General Allotment Act"), and the forcible removal of Native children from their families to faraway boarding schools where their Native practices and languages were degraded and forbidden;
  Whereas officials of the Federal Government and private United States citizens harmed Native Peoples by the unlawful acquisition of recognized tribal land and the theft of tribal resources and assets from recognized tribal land;
  ....

SECTION 1. RESOLUTION OF APOLOGY TO NATIVE PEOPLES OF THE UNITED STATES.

a ACKNOWLEDGMENT AND APOLOGY.—The United States, acting through Congress—

1 recognizes the special legal and political relationship Indian tribes have with the United States and the solemn covenant with the land we share;
2 commends and honors Native Peoples for the thousands of years that they have stewarded and protected this land;
3 recognizes that there have been years of official depredations, ill-conceived policies, and the breaking of covenants by the Federal Government regarding Indian tribes;

4 apologizes on behalf of the people of the United States to all Native Peoples for the many instances of violence, maltreatment, and neglect inflicted on Native Peoples by citizens of the United States;

5 expresses its regret for the ramifications of former wrongs and its commitment to build on the positive relationships of the past and present to move toward a brighter future where all the people of this land live reconciled as brothers and sisters, and harmoniously steward and protect this land together;

6 urges the President to acknowledge the wrongs of the United States against Indian tribes in the history of the United States in order to bring healing to this land; and

7 commends the State governments that have begun reconciliation efforts with recognized Indian tribes located in their boundaries and encourages all State governments similarly to work toward reconciling relationships with Indian tribes within their boundaries.

b DISCLAIMER.—**Nothing in this Joint Resolution—**

1 **authorizes or supports any claim against the United States; or**
2 **serves as a settlement of any claim against the United States.**

(S.J. Res.14 - A Joint Resolution to Acknowledge a Long History of Official Depredations and Ill-Conceived Policies by the Federal Government Regarding Indian Tribes and Offer an Apology to All Native Peoples on Behalf of the United States, 2009)

The significance of the Keepseagle settlement involving Native American farmers with the U.S.D.A. must be understood as different than the Pigford settlement involving Black farmers because of the different historical relationships between Black people and the United States government and Native Americans and the United States government which is ensconced in United States policies. Both groups faced similar discrimination at the hands of the U.S.D.A., especially the local county agencies, however, their settlement process was different though the Keepseagle settlement followed the momentum of the Pigford settlement and was just as substantial. Further, the settlement which also allowed for loan forgiveness such as Pigford, signaled a shift in the relationship between the U.S.D.A. FSA at the federal and local levels in relationship to both Native American farmers and the tribal governments that control some aspect of land usage as a part of sovereign nation territory. And through their process, we also come to understand the concept of tribal food sovereignty that was borne out of the struggles of these farmers. Food sovereignty is the ability to determine and control what type of food best serves you, your family and your community. Tribal food sovereignty understands that the decision on what type of food grown best serves a tribal nation is determined by that specific governing body, in consideration of how decisions are made to best serve their interest.

In fact, according to the BIA (Bureau of Indian Affairs), in 1975 the gross value of agricultural products grown on Indian range and croplands was $394 million. However, Indians received only $123 million or less than one-third of that amount, with the remaining $231 million going to non-Indians. In November 1999, led by Marilyn and George Keepseagle, 213 Indian farmers and ranchers (on behalf of nineteen thousand Indians) filed a $19 billion class-action lawsuit against the Department of Agriculture, alleging a twenty-year history of discrimination in granting of federal loans. As Tex Hall, one of the Indian plaintiffs said, "as indigenous people we are the first farmers and ranchers of this land. All we wanted is a fair chance to become successful."

(Wilkins and Stark, 2017)

The Keepseagle v. Vilsack (Tom Vilsack was the USDA Secretary of Agriculture at the time of the settlement) alleged that the U.S.D.A. discriminated against Native Americans in its farm loan and farm servicing programs. Those who were defined as part of the class settlement included "all persons who are Native Americans farmers and ranchers who (1) farmed or ranched or attempted to farm or ranch between January 1, 1981 and November 24, 1999; (2) applied to the U.S.D.A. for participation in a farm loan program during that same time period; and (3) during the same time period filed a discrimination complaint with the U.S.D.A. either individually or through a representative with regard to alleged discrimination that occurred during the same time period (Bennett, 2010) The individuals who served as representatives of this class included Luke Crasco (Montana – deceased), Gene Cadotte (deceased – buried in Standing Rock Veterans Cemetery – ran the family cattle operation at ranch near Wakpala, South Dakota after serving in the military, Keith Mandan, (Ankara) Porter Holder (Choctaw – Oklahoma), George and Marilyn Keepseagle (Lakota – North Dakota) Claryca Mandan (North Dakota), John Fredericks, Jr., (deceased – Wyoming) and Basile Alkire (deceased – Oklahoma). An article in the Billings Gazette at the time around the final settlement mentioned "The list of named plaintiffs included 142 from Montana and another two from Wyoming "Chinook nation, further stating that the 2007 United States Census of Agriculture indicated 55, 880 Indian operators of which 2,359 were located in Montana" (Thackeray, 2010).

Similar to Pigford, there was also a process of loan forgiveness as part of the Keepseagle settlement that included the following specific language as indicated in court documents:

> Farm Loan Obligation encompasses only direct operating loans, direct farm ownership loans, emergency loans, economic emergency loans, and all amounts due thereunder, including principal, interest, penalties, and charges and debts structured through any Part 766 (formerly Part 1951-S) loan or other farm loan program servicing options.

And because the identifier "Native American" has a specific legal definition that is legislatively determined the settlement specified the following:

1 any citizen of the United States, United States non-citizen national, or a qualified alien who is enrolled in any Indian tribe, band, nation, or other organized group or community, including any Alaska Native village or regional or village corporation as defined in or established pursuant to the Alaska Native Claims Settlement Act (85 Stat. 688) (43 U.S.D. 1601 et seq.), which is recognized as eligible for the special programs and services provided by the United States to Indians because of their status as Indians; or

2 any citizen of the United States, United States non-citizen national or a qualified alien who is enrolled in any Indian group that has been formally recognized as an Indian tribe by a State legislature or by a State commission or similar organization legislatively vested with State tribal recognition authority; or

3 any citizen of the United States who is enrolled in any Indian tribe or "Native group" according to 43 U.S.D. 1602 © and (d) that had an open letter of intent to petition the United States for Federal recognition; or

4 any citizen of the United States, United States non-citizen national, or a qualified alien who can show that, prior to

November 24, 1999, he/she identified himself/herself as Native American. Such self-identification may be established (a) b documentation such as an application for loan or loan servicing assistance submitted to USDA prior to or within the Class Period, or (b) through a credible written narrative, submitted under penalty of perjury, in which the individual describes in detail the circumstances establishing his/her Native American ancestry sufficient to persuade the Track A or Track B Neutral of its genuineness and authenticity. Such a narrative can include recounting a prior instance in which the claimant identified himself/herself as Native American to a governmental entity, such as the U.S. Census Bureau.

Membership in an Indian tribe for purposes of paragraphs (1) – (3), above, shall be defined by the law or rules of the Indian tribe in which the individual claims to be a member. Such membership can be demonstrated by providing a copy of an official tribal document that states that the individual is a member of an Indian tribe, such as (a) an identification card that states that the person is currently an enrolled member of the Indian tribe, or (b) a letter or statement from the tribal government that states that the person is regarded as a member of the Indian tribe.

It is important to understand the particulars of this language and how significant this settlement was from a historical perspective considering the ongoing challenges of tribal governments, Native Americans, community relationship with county governments and their sovereign relations with the United States government. The details of Keepseagle were noted as a $760 million settlement with the U.S.D.A., which allowed for a similar process as Pigford for those who filed claims, though it was handled more expeditiously through the Department of Justice. Specifically, Native American farmers and ranchers were eligible for a payment of up to $50,000 or more and forgiveness of some or all of their outstanding U.S.D.A. loans if they applied or attempted to apply for a farm loan or loan servicing from the U.S.D.A. between January 1, 1981, and November 24, 1999. As with the Pigford I and II settlement process, the $50,000 was under a Track A, which included a 25% additional amount to cover taxes expected to be paid on this amount to the Internal Revenue Service (I.R.S.). The Track A was for those who did not have all of the necessary documents to prove that the specific level of discrimination that they had suffered.

If the Native American farmer or rancher had more specific proof of the discrimination they had suffered at the hands of the U.S.D.A., they could receive up to $250,000 under what was designated as Track B, which also included an additional amount necessary to cover the taxes expected to be paid to the I.R.S. The U.S.D.A. has also agreed to make some changes to its farm loan programs to help make sure that these programs meet the needs of Native American farmers and ranchers. The loan forgiveness is not part of the $760 million but comes out of a second fund created for class members that provides up to $80 million for full or partial loan forgiveness with the U.S.D.A. paying up to an additional $20 million for the costs of administering the settlement (Bennett, 2010).

Because they did not want any remaining money from the settlement to simply return to the United States government, they allowed for it to be donated to any organizations that specifically provide agricultural, business assistance or advocacy services to Native Americans. The Settlement additionally provides for improvements in the delivery of Farm Loan Program services to Native Americans. There was a specific Native American Farming and Ranching Council created to ensure that the farm loan program was responsive to the particular needs of Native American and Alaska Native farmers and ranchers in partnership with the FSA and U.S.D.A. There was also a specific governmental U.S.D.A. Ombudsperson position created for Native American and other SDA farmers and ranchers to serve as a point of contact to specifically track civil rights complaints or any challenges with accessing programmatic support. At the last meeting this council held which was in May of 2018 this was reflected in their minutes:

- Supplemental award checks mailed to almost 3600 individuals – prevailing claimants during the claims phase of Keepseagle case;
- $18,500 direct and $2775 to I.R.S. to offset taxes as a result of the settlement;
- $38 million to support non-profit organizations that support Native American farmers and agriculture;
- $268,000 remaining placed in the Native American Agriculture Fund.

And last as part of the settlement agreement, the U.S.D.A. established 10–15 regional venues to provide technical training and support and developing a plain language guide to the application for farm loans and loan servicing. If funds are available, the U.S.D.A. will also fund and staff consolidated sub-offices at Tribal Headquarters on Indian Reservations.

"Earlier this month the USDA agreed to pay out $760 million to thousands of Native American farmers as part of a historic settlement of a separate class action discrimination lawsuit that claimed they were denied access to low interest government loans available to white farmers. But Native American farmers will not have to wait for Congress to allocate the money. It will come from a fund administered by the Department of Justice to pay judgements against the United States."

(Sturgis, 2010)

There has been an ongoing struggle in the relationship between the United States government and tribal nations that has been tied to the long history of colonialism and the manner in which the over 400 treaties that were signed have been either ignored or only marginally recognized. The process of removal meant that many Native Americans were forced to change some of their traditional practices including the type of food they grew and the manner in which it was produced. Their displacement was not just based on the acquisition of their land, it had the unfortunate consequences of disrupting their food system and means of survival.

In their book, "American Indian Politics and the American Political System," Wilkins and Stark provide an overview of the general Native American agricultural terrain detailing the struggle of these nations to succeed due to the control of the Bureau of Indian Affairs through the Department of Interior of their land which is held in trust funds. Today, they offer that

The Department of the Interior currently manages about fifty-six million acres of Indian trust land throughout Indian Country. It administers more than one hundred thousand leases and $3.5 billion in trust funds. For example, in 2009 the Interior collected about $298 million for more than 384,000 open IIM accounts. In addition, $566 million was collected for about 2,700 Tribal accounts for more than 250 tribes. Untold billions have flowed through both accounts since the 1880s, and neither individual Indians nor Tribal governments have ever received a thorough accounting of their moneys. In fact, the Indian plaintiffs assert that they have lost upward of $137.5 billion since 1887.

(Wilkins and Stark, 2017)

They go on to state that on Indian lands there are three major types of agriculture practiced that include livestock grazing, dry land farming and irrigation. "Indian farmers typically have concentrated their energy and resources on grazing which tends to be the least profitable type of farming" (Wilkins and Stark, 2017). Native American farmers, like other marginalized farmers who have been put into the position of seeking redress from the United States government, have used unique and creative measures to simply survive. They adopted techniques because they best served the purposes for the conditions in which they were forced, but they also faced many environmental conditions that forced them to simply make choices as vehicles for survival.

References

Bennett, D. (2010). *Keepseagle discrimination settlement: $680 million.* Clarksdale, MI: Western Farm Press http://search.proquest.com/docview/847388773/abstract/561F158B9F9E4AD9PQ/19.

Cobell v. Salazar. (2009). U.S. Department of the Interior.

Dunbar-Ortiz, R. (2014). *An Indigenous peoples' history of the United States.* Boston, MA: Beacon Press.

Garvey, T. (2010). *The Indian Trust fund litigation: An overview of Cobell v. Salazar.* Congressional Research Service.

Ives, S. (1996). *The west: The geography of hope* [Documentary].

Nishime, L. & Williams, K. D. H. (eds). (2018). *Racial ecologies.* Seattle, WA: University of Washington Press.

S.J. Res.14—A joint resolution to acknowledge a long history of official depredations and ill-conceived policies by the Federal Government regarding Indian tribes and offer an apology to all Native Peoples on behalf of the United States. (2009). 111th Congress (2009–2010). https://www.congress.gov/bill/111th-congress/senate-joint-resolution/14/text.

Sturgis, S. (2010). *"Still waiting for justice, Black farmers rally in North Carolina.* Durham, NC: Institute for Southern Studies.

Thackeray, L. (2010, November 13). Discrimination settlement cold mean payments for Montana Indian farmers. *Billings Gazette.*

Wilkins, David and Heidi Kiiwetinepinesiik Stark. (2017). *American Indian Politics and the American Political System.* Spectrum Series.

6   The marriage of the Garcia and Love settlements

In November 2012, National Public Radio did a feature on Rosemary Love. They reported that on February 1, 1983, Rosemary Love was visited by the head of a local U.S.D.A. office while she was in the hospital recovering from surgery for breast cancer. She was a farmer in Montana and had taken out loans but was behind on her payments. She said he wanted her to sign a new security agreement but she refused because it would give the U.S.D.A. claim to her property, her livestock and the next season's crops. Later that year, the U.S.D.A. accelerated her loan, demanding payment in full in 30 days so she was forced to declare bankruptcy which leads to the government seizing portions of her family's ranch and having it liquidated. But she found out some years later that other ranchers in that area, all male, had received loan forgiveness, which she ended up testifying about before Congress (Robinson, 2012).

The settlement process governing both Hispanic/Latino farmers (Garcia) and women farmers (Love) followed the settlements of the Black farmers (Pigford) and Native American farmers (Keepseagle), however, they were handled concurrently with one based on race/ethnicity and the other on gender. Considering the intersectionality of the women farmers, we must acknowledge that these women could also have been covered as a member of the other groups or at least been a family member of a recipient yet not included as a principal operator.

This chapter will discuss both the case of the Hispanic/Latino(a) farmers (Garcia) and the women farmers (Love) because they had similar outcomes in that, while their ability to rise to the level of a class-action suit similar to the Black farmers and Native American farmers was ultimately negated after many attempts, the U.S.D.A. did enter into a voluntary settlement process for both groups. And as with the other settlements, changes at the Department of Agriculture with regard to outreach practices with women and Hispanic/Latino farmers became more supportive as a result of the settlements. Because their settlement process was entered into from a voluntary standpoint, after many attempts at a class-action lawsuit, the amount was not as substantial, but for women especially at least their profile as farmers was elevated. The lawsuit Love v. Vilsack was filed in the U.S. District Court for the District of Columbia initially on October 19, 2000, by a number of women farmers against the U.S.D.A. for gender discrimination in the administration of the U.S.D.A. farm loan programs. The women farmers first sought to have the case certified as a class action on behalf of women farmers and prospective women farmers nationwide who were discriminated against in attempting to obtain farm loans or loan servicing from the Farm Services Agency (F.S.A.) from 1981 to December 31, 1996 and from October 19, 1998 to present (Bennett, 2011).

The United States Department of Justice (U.S.D.o.J.) confirmed patterns of civil rights violations at the U.S.D.A. against women and minority farmers, and a few women even testified that they were offered loans in exchange for sexual favors. Ultimately, a voluntary claims process of settlement was put in place but by the time this occurred a significant level of damage and loss had already been suffered by women farmers (Robinson, 2012).

The Census of Agriculture only began asking for the gender of principal farm operators in 1978. They note that prior to 2002 the data collected would only account for one operator per farm, which largely eliminated women if they were part of a husband/wife team specifically. Subsequent to that 2002 Ag Census, all operators are counted, up to three operators per farm, which then allows for a more comprehensive approach to understanding who is involved in these farm operations. The "principal" operator is noted as the person who is in charge of the day-to-day decision-making for the farm or ranch (*USDA – National Agricultural Statistics Service – Census of Agriculture*, n.d.).

The 2007 Census of Agriculture notes that the states with the highest percentage of female principal farm operators. In the United States, at that time were Arizona (38.5 percent), New Hampshire (29.7 percent), Massachusetts (28.9 percent), Maine (25.1 percent) and Alaska (24.5 percent). The states with the lowest percentages of women principal operators are in the Midwest. The share of farms operated by women nearly tripled over the past three decades from five percent in 1978 to 14 percent in 2007, according to the U.S. Department of Agriculture Census. Looking at the actual number, between 1982 and 2007 the number of women-operated farms grew from 121,600 to 306,200 – with increases in all sale classes. A majority of women-operated farms have had annual sales of less than $10,000, which is where most of the growth of women-operated farms has occurred (*2007 Census of Agriculture – History*, 2007).

However, while there is an increase in the number of women farmers most of them are part-time farmers with other full-time jobs to maintain stability. This has been a growing trend with a number of younger farmers, regardless of gender, who have increasingly relied on social media to maximize their ability to sustain their productivity while not relying solely on farming as a means of support, particularly at the beginning stages.

Five percent of women-operated farms (15,400) had sales of $100,000 or more in 2007. Most of these farms specialized in grains and oilseeds, specialty crops, poultry and eggs, beef cattle and dairy. The poultry and egg specialization alone accounted for roughly half of the women-operated farms with sales of one million dollars or more. Nearly half of the farms operated by women specialized in grazing livestock, and if we count the women who are secondary operators, the number of women farmers increases to one million (*2007 Census of Agriculture - History*, 2007).

A racial and ethnic breakdown of women farmers indicates that a significant majority are white, followed by American Indian/Alaska Native Farmers, Spanish, Hispanic or Latina farmers, Black or African-American, Asian and then Native Hawaiian or Other Pacific Islander farmers. It is important to also note that the U.S.D.A. maintains a separate Census of Agriculture for American Indian Reservations (*USDA – National Agricultural Statistics Service – Census of Agriculture*, n.d.)

According to the 2012 Census of Agriculture, there was a 2 percent decrease since the 2007 agriculture census, though nationally women maintained their status as 30 percent of farmers. Of total female farmers, 288,264 were principal operators, that is, the person in charge of the farm's day-to-day operations. It notes that Texas had the most farmers; however, Arizona had the highest proportion of women farmers (45 percent of all farmers in the state.) As principal operators, the number of women decreased six percent with a decline of four percent of total farms. Another interesting fact from the 2012 Census of Agriculture that will be key to compare with the most recent agriculture census is that women principal operators at that time were noted as older than principal operators overall. Only four percent of women principal operators were under 35 years old compared to six percent overall, with the average age of women 60.1 years. However, women principal operators increased their use of technology but, as stated previously, are less likely to report farming as their primary occupation (*USDA – National Agricultural Statistics Service – Census of Agriculture*, n.d.).

Looking at the 2012 agriculture census noting farms and sales, women principal operators sold $12.9 billion in agriculture products including $6 billion in crop sales and $6.9 billion in livestock sales. They operated 62.7 million acres of farmland and sales by women represented 3.3 percent of total U.S. agriculture sales. A total of 6.9 percent of United States farmland was operated by women. Eighty-two percent of farms with a woman principal operator had fewer than 180 acres and 76 percent had sales of less than $10,000 in 2012 (*USDA – National Agricultural Statistics Service – Census of Agriculture*, n.d.)

The 2017 Census of Agriculture, the latest to date indicates a 27 percent increase of women farmers since the last census which is large because of the revised data collection process which asks about all that are involved in the decision-making process with the farm. This census indicated that there are 1.2 million female producers, which is 36 percent of the country's 3.4 million total producers. It also mentions that they are slightly younger and more likely to be a beginning farmer and yet more than half of all farms (56 percent) had female producers. Also, female-operated farms sold $148 billion in agricultural products in 2017, with 49 percent (73 billion) in crop sales and 51 percent (75 billion) in sales of livestock and livestock products (*Female Producers*, n.d.) It is an important fact that the determination of who legally constitutes a farmer has become more inclusive of varying relationships to the farm itself. This allows for greater access to U.S.D.A. resources and support and provides for a greater understanding of the full capacity of the agricultural arena that includes both large and small farms. Ironically, one aspect of the negation of the women farmers as a class-action lawsuit against the U.S.D.A. along with the Hispanic/Latino farmers was the difficulty in determining the size of the group that would be included as part of the "class status" (Bennett, 2011).

Women are organizing to grow their agricultural businesses and to make sure that these discriminatory practices don't happen again. For example, there is the board membership for the top five National Commodity Organizations that entail the board gender breakdown as follows – 2010/2011 figures:

1 National Corn Growers Association – Males – 14/Females – 1.
2 American Soybean Association – Males 45/Females – 1.

3 National Association of Wheat Growers – Males – 51/Females – 1.
4 National Cotton Council – Males – 92/Females – 0.
5 United States Rice Producers Association – Males – 23/Females – 0.

However, the leaders of all three national organic food and agriculture organizations are women. The Organic Center is lead by Jessica Shade – Director of Science Programs. The Organic Trade Association is led by Christine Bushway, and the Organic Farming Research Foundation is led by Maureen Wilmot. Twenty-two percent of organic farmers are women who follow practices that conserve soil and biological diversity by rotating crops and avoiding synthetic fertilizers, pesticides, hormones and genetically modified seed. Critical networking organizations focusing on women farmers also include the Women Food & Agriculture Network (W.F.A.N.) which came together from a working group that prepared for the United Nations Fourth World women's Conference held in Beijing, China. They have programs such as the Plate-to-Politics which trains women in agriculture to get involved in the political arena so that their voices can be heard. And at the W.F.A.N. Conference in Des Moines, Iowa that I attended, they even featured a vendor who sold farming clothes that were specifically made for women farmers (See photo 6.1).
Other programs include:

1 Harvesting Our Potential – a mentorship program or new and aspiring women farmers.
2 The Outreach Program of the U.S.D.A. F.S.A.
3 American Agri-Women – the nation's largest coalition of farm, ranch and agribusiness women which was actually founded in 1974.
4 National Women in Agriculture Association – a faith-based minority agriculture organization.

"In Texas and areas throughout the Southwest, Hispanic farmers also struggled against years of USDA indifference and discrimination in the departments farm credit loan programs disaster relief loans and other aid and benefit programs. By denying minority farmers loans, disaster assistance and other aid frequently given to whites, in essence the USDA policies helped to drive minority farmers out of business, resulting in the loss of minority farms and lands."

(Evans, n.d.)

Photo 6.1 Women Food and Agriculture Network hat and scarf.

A group of 110 Hispanic farmers brought a lawsuit against the U.S.D.A. for some patterns of discrimination. The Garcia v. Vilsack lawsuit involved allegations that U.S.D.A. unlawfully discriminated against Hispanic/Latino farmers. This lawsuit was named after Guadalupe L. Garcia who was a third-generation farmer from Las Cruces, New Mexico. In an article in *High Country* News, Garcia is quoted as stating "To add insult to injury, F.S.A. assisted the Anglo farmers in purchasing our farms at a special master's sale," he wrote. "In fact, one of the purchasers was the same neighbor who years earlier had stated that it would only be a matter of time before he would own our farm" (Waddell, 2020).

Specifically, the lawsuit, which was filed in the U.S. District Court for the District of Columbia in 2000 on behalf of all similarly situated Hispanic farmers across the country, alleged that U.S.D.A. discriminated against them with respect to credit transactions and disaster benefits and suit further claimed that U.S.D.A. systematically failed to investigate complaints of discrimination, as required by U.S.D.A. regulations (Feder and Cowan, 2013).

> It was in the 2002 ruling where the district court considered the Hispanic farmers' motion for class action status. The Federal Rules of Civil Procedure authorize class action lawsuits, in which one or more individuals are allowed to sue on behalf of all members of a class under certain circumstances. Motions for class action status are reviewed by the courts, and parties seeking class certification must show, among other things, that (1) the class is so numerous that joinder of all members is impracticable, (2) there are questions of law or fact common to the class, (3) the claims or defenses of the representative parties are typical of the claims or defenses of the class, and (4) the representative parties will fairly and adequately protect the interests of the class. Ultimately, the district court in Garcia denied the Hispanic farmers' motion for certification of a class consisting of [a]ll Hispanic farmers and ranchers who farmed or ranched or attempted to do so and who were discriminated against on the basis of national origin or ethnicity in obtaining loans, including the servicing and continuation of loans, or in participating in disaster benefit programs administered in the United States Department of Agriculture, during the period from January 1, 1981 through December 31, 1996, and timely complained about such treatment, or who experienced such discrimination from the period of October 13, 1998 through the present.
>
> (Feder and Cowan, 2013).

In particular, the fact that multiple U.S.D.A. employees in multiple jurisdictions were responsible for making eligibility decisions made it difficult for the farmers to establish that there was a common policy of discrimination. The U.S.D.A. had a variety of reasons for denying loans, including credit information and farming experience, meant that the farmers could not point to a common facially neutral U.S.D.A. policy that had led to a statistically relevant racial imbalance in the denial of loans (Feder and Cowan, 2013).

Because many of the claimants may not have the means to pursue litigation on their own and because other Hispanic farmers who were not a party to the litigation but who may have been victims of discrimination might have missed the statute of limitations for filing some of these farmers have also pressed members of the executive and legislative branches to help them resolve the case and secure compensation. Such efforts intensified in the wake of the settlement agreements D.o.J. entered into with Native American farmers in Keepseagle and with the second group of Black farmers in the case commonly referred to as Pigford II, and D.o.J. eventually made a settlement offer in the Garcia case.

The U.S.D.A. in conjunction with D.o.J. established a voluntary process to settle the claims of Hispanic and female farmers in 2011. Under the settlement, $1.33 billion is available to compensate eligible farmers for their discrimination claims, as well as an additional $160 million in debt relief. Awards of up to $50,000 or $250,000 are available, depending on the type of claim and evidence submitted, and successful claimants may also be eligible for tax relief and loan forgiveness. It is important to note that the Garcia and Love settlement process was voluntary, though individual women and Hispanic/Latino farmers did have a choice that allowed them the ability to deny this process and pursue other avenues (Feder and Cowan, 2013).

The 2017 Census of Agriculture indicated 112,451 Hispanic/Latino producers which it says accounted for 3.3 percent of the country's 3.4 million producers. The majority of Hispanic producers are in Texas, California, New Mexico and Florida. Between the 2012 and 2017 Census of Agriculture, there was an eight percent increase, 86,278 farms. The majority of Hispanic/Latino farmers resides in Texas and account for ten percent of the state's total producers. In terms of a larger share the census notes that in New Mexico, 30 percent of all producers identified as Hispanic. Hispanic-operated farms accounted for 32 million acres of farmland, 3.6 percent of the U.S. total and the majority (61 percent) of these farms were less than 50 acres (*Hispanic Producers*, n.d.).

References

2007 Census of Agriculture—History (Volume 2, Subject Series Part 7). (2007). United States Department of Agriculture/National Agriculture Statistics Service. https://www.census.gov/history/pdf/2007aghistory.pdf.

Bennett, D. (2011). *Love v Vilsack: Women farmers allege USDA discrimination.* Southwest Farm Press; Clarksdale. http://search.proquest.com/docview/874476157/abstract/7A8D19FC69824D7CPQ/1.

Evans, D. (n.d.). *Hispanic farmers seek justice for discriminatory USDA lending practices.* Durham, NC: Institute for Southern Studies.

Feder, J. and Cowan, T. (2013). *Garcia v. Vilsack: A policy and legal analysis of a USDA discrimination case.* Congressional Research Service.

Female Producers. (n.d.). 2017 Census of Agriculture.

Hispanic Producers. (n.d.).

Robinson, J. (2012, November 9). *Women, hispanic farmers say discrimination continues in settlement.*

USDA – National Agricultural Statistics Service—Census of Agriculture. (n.d.). Retrieved February 19, 2019, from https://www.nass.usda.gov/AgCensus/.

Waddell, B. (2020, March 11). Men like you weren't meant to own land. *High Country News.*

7 Envisioning a more open U.S.D.A. for the greater good

There are a few things that the individuals covered by the Pigford, Keepseagle, Garcia and Love settlements have in common. The Black, Native American, Latino/Hispanic and women farmers, as indicated by the settlements did not receive the support they needed to be successful. Each one of the groups suffered tremendous land loss and they all relied to a certain extent on their agrarian roots and traditions, to create mechanisms for survival. And it is the fact that these groups persevered and challenged the U.S.D.A. to change and to develop better outreach policies that have also contributed to the food justice movement from a governmental position because they had to focus on the capacity of the work of these farming groups which are so necessary toward addressing food deserts. While acknowledging that we are a part of a global food system, the solution to healthier food access is often concentrated at the local level.

While it was important to understand the historical landscapes and backgrounds that led up to each settlement, as well as to know the details of the settlements for each of the groups, some of the strategies and approaches to farming can be directly traced to the struggles of these groups. And it is the strategies and approaches of their survival that can be categorized as being a part of a food justice movement because their goals were to navigate an unjust system and still survive. Additionally, where the work of the U.S.D.A., especially at the county level was targeting individual farmers, the Black, Native American, Hispanic/Latino and women farmers took collective approaches to supporting one another and bolstering the capacity of these mostly small farms through pooling their resources.

First, looking at the issue of land, there has been a decrease in the number of acres that make up many smaller family farms while large more agribusiness farms have increased in size. An increase in the number of farmers that operate as tenant farmers has taken place though these arrangements do not mirror the tenant and sharecropper agreements that were so exploitive of Black people in the 1920s–1950s period. Many young farmers have also engaged in techniques that allow them to maximize the usage of small acres of land in ways where they manage to grow more produce. The most recent Census of Agriculture also accounts for the use of technology of farmers which is fairly common among young farmers to use technology to explore the market for their goods. Regarding land ownership Eric Holt-Gimenez writes the following:

> "Over the past three decades, shifts in ownership and increases in farm size have seen more renters (farmers who rent 100 percent of the land they farm) and part-owners farming a growing number of acres, especially in the agriculturally productive Midwestern United States…In 2012, agricultural producers rented and farmed nearly 354 million acres of farmland, nearly 40 percent of total U.S. farmland, according to the results of the USDA's Tenure, Ownership, and Transition of Agricultural Land (TOTAL) survey. Of this land, individual farmers own 20 percent, while the remaining 80 percent is rented out by non-farming landlords, either as individuals or participants in differing ownership arrangements."
>
> (Holt-Gimenez, 2017)

While I am suggesting that food justice efforts such as farmers' markets, Community-Supported Agriculture (C.S.As.), and farming cooperatives and collectives serve an important purpose in addressing issues of food deserts and economic conditions that affect access to healthy food, these efforts are difficult to sustain and often do require buy-in from the local community that they are serving, as well as governmental support. In some of the examples, I will offer in this chapter another commonality surfaces and that is an alignment of food justice initiatives with other areas of injustice that are being addressed through efforts of schools, churches and other non-governmental organizations, though access to food is at the center. The proximity of these initiatives to those locations allows for a seamless engagement, that is, organic in development allows for greater participation than initially anticipated. This is how farmers' markets operate and during the course of my research for this book, I visited several farmers' markets in a number of different states. In fact, whenever I travel I look for a farmers' market to provide me a sense of the connections and life of the community and to see what type of fresh produce is available, along with other products.

I also consider the location of the farmers' market in terms of proximity to public transportation. Last, I look to see if the patrons of the market are diverse, depending on the location, the costs of the food, and with regard to the U.S.D.A., I look to see if there is an accommodation for Supplemental Nutrition Assistance Program (S.N.A.P.) recipients, such as the ability to use their E.B.T. funds to make purchases. On their website, the U.S.D.A. provides information on their partnerships with farmers' markets on how to double the use of E.B.T. funds, how to make healthy choices and even provides a National Directory of Farmers' Markets (Using SNAP at Farmers' Markets n.d.).

On a Saturday morning in the summer a couple of years ago, I went to the farmers' market in Brooklyn, New York, which is located in a more higher-priced area of Brooklyn, as housing prices go in New York which tends to be one of the more expensive places to live in the United States, but it was fairly easily accessible via public transportation. New York has an extensive subway system, and the farmers' market was about two walking blocks, and through a park from the closest subway station. This Brooklyn Farmers' Market was also located right next door to the Brooklyn Public Library (B.P.L.), which coincidentally had a popular reading program taking place in the Children's area of the B.P.L. at the same time. This location was also about a half-mile walk from the Brooklyn Art Museum, which happened to have a vibrant First Friday Afro-Caribbean focus with bands and food and a Frida Kahlo exhibit that I visited. I did not see any signs regarding S.N.A.P. benefits at this market, but I do see them at the farmers' markets I frequent year-round where I live. However, I did see a lot of farmers and other vendors from diverse backgrounds and in addition to a lot of fresh produce, some indicated as "organic," there was quite a variety of additional food products available such as gluten-free baked goods. Because of the location of the farmers market in Brooklyn, some of the farmers actually traveled there from further upstate New York or neighboring states. Given the movement of food in our global system, a regional connection can still be considered fairly local.

Within a 20-mile range of where I live in Indiana I have access to at least five farmers' markets that offer a variety of seasonal fruits and vegetables grown relatively locally, and other types of healthy and nutritious food options. While that range is considered outside of the guidelines of the U.S.D.A. in what determines whether or not you live in a food desert, which I do, it is because I have access to my own transportation that allows me to have and make better choices in terms of my own diet and nutrition. And at each of those five farmers'

Photos 7.1 and 7.2 Author at Brooklyn NY Farmer's Market and additional Brooklyn Farmer's Market photo.

Photo 7.3 Brooklyn evolutionary organic farmer's market.

markets, I have seen signs specifying the S.N.A.P./E.B.T. benefits with specific people at the market "information booth," who are helpful in explaining the process. There are also signs, especially at the booths of the mostly farmer vendors at the market who indicate their openness to people shopping with S.N.A.P./E.B.T. benefits – which is a majority of the vendors, depending on what they are selling.

Many raise questions regarding the inclusion of farmers' markets as part of the food justice movement and rightfully so when it comes to true accessibility and associated costs. Some present as boutique options available to communities that would not be considered food deserts and that have no proximity to neighborhoods that would fit that category. And yet many, like the ones that I frequent and others I have visited in less affluent communities do provide access to healthy options at affordable prices and are addressing some level of disparities in our localized food system (Alkon, 2012).

In a report sponsored by the Union of Concerned Scientists, they refer to a report by The Farmers Market Coalition who notes that the number of farmers' markets actually doubled between 2000 and 2010 to over 6100 (O'Hara, 2011). According to the U.S.D.A., there are

over 8600 farmers' markets today that are part of their national directory and you can even place your zip code into their site to determine
how many are within a certain 5, 10, 20, 50 or100 mile radius of where
you live. I actually have 10 farmers' markets within a 10-mile radius
of my home (Using SNAP at Farmers' Markets n.d.). Following are
some policy changes that are recommended by the Farmers Market
Coalition that, while not necessarily speaking deliberately to a notion
of justice, do provide a food system context for the place of these farmers markets:

*The Farmers Market Coalition recommended the following policy
changes:*

1 *Increase funding for programs that support local and regional food
   systems.*
2 *Raise the level of research on the impacts of local and regional food
   systems.*
3 *Restructure the safety net and ensure credit accessibility for local
   food system farms.*
4 *Foster local capacity to help implement local and regional food system plans. Support the realization of farmers' market certification
   standards* (O'Hara, 2011)

Historically, farmers' markets are stated to have begun during the
colonial period where buying local and fresh produce, meats and baked
goods were only available at various open markets in the communities.
"In the first decades of this century (20ᵗʰ Century), most cities with at
least 30,000 people sponsored municipal markets. But better roads and
refrigeration soon ushered in supermarkets and wholesalers, leaving
many small farms, and markets out of the food cycle" (Jablow and
Horne, 1999). However, in terms of rural environments, and especially
in rural Black Belt communities of the South, many of the local farmers would set up mid-week in the town square where they could sell
their products directly to the local community. This still occurs today
and it has happened a lot where there were no grocery stores easily
accessible and so farmers' markets and similarly farm stands would
and still do fill the gap.

There is another aspect to farmers' markets that also fits into the
arena of food justice and that is the solidarity that is established across
people, especially beyond the rural/urban/suburban divides, among
other constructed but very real differences. Within the simple exchange
of goods and currency are the connections made, especially if there is

a mutual reliance established that speaks to the ability of the small farmer to be sustained through the sale of their products, and the ability of the consumer or patron to purchase healthy, locally grown foods for the intrinsic benefit. There becomes awareness between them as individuals but also both as part of shared communal experiences and they develop an awareness of their mutuality as experienced within this global food system that relies upon this local exchange.

"Local and regional food systems create jobs and raise incomes in the areas they serve, keeping customers food dollars active in the local economy as farmers increase spending on inputs and equipment to meet growing demand...Food sold through direct marketing channels tends to be relatively less processed, which can result in significant energy savings – so expansion of local and regional food systems can also reduce environmental impact."

(O'Hara, 2011)

Our reliance upon these localized exchanges as far as food is concerned is increasing, as evidenced by the growth in farmers' markets along with the related changes in the global economy that require us all to think more creatively about what we should all expect to have access to from a justice standpoint.

Other farming practices that fit within a food justice framework include farm-to-table movements that are the focus of many restaurant and café offerings as well as many schools and C.S.As. which, like farmers' markets, have an associated cost that make the justice aspect not necessarily automatic. There are so many different types across the country, some that are very specialized in providing particular options like vegan/vegetarian or an even more narrowly defined specific food type – like greens only – or just eggs. The key to most of them is that they still are focused on providing locally grown and/or created healthy food to communities through either delivery service or regular weekly or monthly pick-ups. I will mention one to which I personally subscribe because it covers multiple approaches to food justice and also food sovereignty.

There is a youth-based advocacy and community development organization called Kheprw Institute that has a "Community Controlled Food Initiative" (C.C.F.I.) among other projects. Similar to a C.S.A., once a month people in the community can purchase a bag of produce, plus a dozen eggs, some of which is provided by a collective of local Black farmers. Occasionally, they have a healthy food cooking demonstration and there is an option for purchase available for those who use S.N.A.P./E.B.T. funds. This initiative is one that grew out of the closure of the Double 8 grocery store, mentioned previously in the text and it has been in place for a few years, so very sustainable. As author Tabitha Barbour, who conducted research on Kheprw Institute points out,

> Like CCFI there is a growing movement for community controlled food where it provides more than a handout of food in the community but empowers neighborhood residents to find solutions and revitalize their own spaces.
>
> (Barbour, 2018)

The work of Kheprw, Inc. to serve as one vehicle of food distribution for a local Black farming collective fits into an aspect of the food justice movement identified as food sovereignty. There are many urban community farms and also farming collectives who approach the work that they do as having an end goal of not just making healthy food accessible to various marginalized communities but additionally as empowering those same communities to determine what type of is best for them, which may have some specific cultural roots of which they identify.

Women, Food and Agriculture Network (W.F.A.N.) held their annual conference a couple of years ago in Des Moines, Iowa, where participants were invited to an "Urban Agriculture Tour," which featured a

Photo 7.4 Lutheran Church Global Green Refugee Farmer's Collective.

Photo 7.5 Des Moines, IA Homeless Shelter Aquaponics training.

community garden that was next to a pre-school, a neighborhood farm that included a store where local producers could collectively sell their products, a farm incubator for refugees called Global Green Farms that was created by the Lutheran Church and an Aquaponics system as part of a farming initiative at a local homeless shelter. At the heart of all of these urban agriculture, endeavors were the goal of food sovereignty and a commitment of communal investment toward helping people become more than self-sufficient.

The mission of W.F.A.N. is "To engage women in building an ecological and just agricultural food system through individual and community power." Their vision is "a vibrant, community-centered food and agricultural system in which women are strong leaders." According to the history noted on their website, the organization, which is one of the few in the United States that has a social justice focus for women farmers, was "started in 1994 by Iowa organic farmer Denise O'Brien and New York state food justice advocate Kathy Lawrence when they organized women in agriculture working group for the United Nations 4th World Conference on Women held in Beijing" (About WFAN – Women, Food and Agriculture Network n.d.). At the annual conference, there were at least a few hundred women farmers from multiple generations who were all very clear on the values of the organization and the necessity of sharing their farming skills. There was a presentation on no-till farming by two women soil quality experts from the U.S.D.A., a vendor who sold farm clothes specifically designed for women, a presentation on "Dismantling White Supremacy in Our Food Systems Work," and the keynote was Aleysa Fraser, one of the founders of the "Black Dirt Farm Collective" located on the Eastern Shore, 2100-acres of land that was occupied by Harriet Tubman. Among the things that Ms. Fraser said in her presentation was "I returned to the same land looking for freedom," and "Feed the Village, Feed the Revolution." An article in the Atlanta Black Star about the Black Dirt Farm Collective states that it is "all about paying homage to the Black farmers who came before them.

> The group, comprised of 10 farming-collective members, prides itself on producing high-quality, nutrient-rich eggs, fruits, vege-tables, small grains and proteins. The farm's land has long been a part of the freedom struggle, as its revered elder, abolitionist Harriet Tubman, once rescued her parents and nine others from the very same place. To retain Black, in Black DIRT, is to pay homage to the Black agrarian experience and to the totality of the agrarian struggle in the Americas," according to the farm's Facebook page. It works to continue adding to "the contributions that folks of color have made to agriculture and society at large.
>
> (Kenney, 2017)

There is an additional component when you look at food justice and food sovereignty through the lens of the indigenous nations. The defi-nition of tribal food sovereignty is "The right for Indigenous nations to define their own diets and shape food systems that are congruent with their spiritual and cultural values" (Food Sovereignty and the Rights of Indigenous Peoples 2013). Communities that exhibit tribal food sov-ereignty and food sovereignty are those that:

- Have access to healthy food;
- Have foods that are culturally appropriate;
- Grow, gather, hunt and fish in ways that is maintainable over the long term;
- Distribute foods in ways so people get what they need to stay healthy;
- Adequately compensate the people who provide food;
- Utilize tribal treaty rights and uphold policies that ensure contin-ued access to traditional foods (Tribal Food Sovereignty defini-tion n.d.).

It becomes critically apparent that the determination of tribal nations post-Keepseagle is to gain back control over land, farming and water rights that were a part of negotiated treaties with the United States government. The U.S.D.A. has even moved to gather a separate data set of Native American farmers as part of their Census of Agriculture. This works to ensure that the U.S. government maintains the complex relationship that defines the policies contained in treaties and the fact that many of them have been breached with no consequences is another component of injustice that can no longer go unaddressed.

References

"About WFAN - Women, Food and Agriculture Network." *Mission of WFAN.*
Alkon, A. H. 2012. *Black, White, and Green : Farmers markets, race, and the green economy.* Athens, Greece: University of Georgia Press.
Barbour, T. C. 2018. "A community's collective courage: A local food cooperative's impacton food insecurity, community and economic development and climate change." *Butler Journal of Undergraduate Research.* 4.
Food Sovereignty and the Rights of Indigenous Peoples. 2013. International Indian Treaty Council.
Holt-Gimenez, E. 2017. *A foodie's guide to capitalism: Understanding the political economy of what we eat.* New York, NY: Monthly Review Press.
Jablow, V. and Horne, B. 1999. "Farmers' Markets." *Smithsonian Magazine.*
Kenney, T. 2017. "6 Successful Farms Promoting the Resurgence of Black Agrarians." *Atlanta Black Star.* https://atlantablackstar.com/2017/05/30/6-successful-black-owned-farms-promoting-resurgence-black-agrarians/4/.
O'Hara, J. 2011. *Market forces: Creating jobs through public investment in local and regional food systems.* Union of Concerned Scientists.
"Tribal Food Sovereignty Definition."
"Using SNAP at Farmers' Markets." https://snaped.fns.usda.gov/nutrition-education/nutrition-education-materials/farmers-markets.

8 Conclusion

The four settlements of Pigford (Black farmers), Keespeagle (Native American farmers), Garcia (Hispanic/Latino farmers) and Love (women farmers) with the U.S.D.A. went largely unnoticed by the general public. The case of the Black farmers most likely received more publicity than the other settlements because it was the first one with the U.S.D.A. and due to the very visible activism of the group and the relatively high profile of Dr. John Boyd, President of the National Black Farmers Association. He was the only non-white farmer that was part of a recent History Channel series called "The American Farmer."

The discrimination experiences of the Black farmers were not too difficult for the United States public imagination to grasp because their experiences aligned squarely with common historical knowledge of Black history initially shaped from a condition of enslavement, sharecropping and tenant farming, Black Code laws and the subsequent struggle of significant land loss. The Keepseagle settlement brought to life the ongoing problems associated with the United States government both recognizing and honoring the treaty relationships established with Indigenous nations but subsequently recognizing that breaching the trustee relationship associated with tribal land management has led to suffering stemming from the unjust and illegal behavior of the United States government.

The Garcia and Love settlements were only connected to the extent that their settlements were voluntary and, therefore, processed simultaneously. But they were also similarly related because it was a challenge for the government to conceptualize both groups as worthy of a class structure for the purposes of a class-action lawsuit, and further that their lack of success truly happened as a result of deliberate U.S. government discrimination. The average U.S. citizen generalizes and associates Latinos with farmworkers and women are stereotyped as "wife of farmers," who are not themselves also landowners. This partially explains the necessity for the Census of Agriculture to change how principal operators are counted.

> Agricultural land, once a measure of wealth and power and a means by which to produce value is now a financial asset, its value atomized and repackaged, bought, sold, and circulated in global markets at the speed of a keystroke. The land, of course, never moves, but its ownership changes rapidly. Rents produce a steady income stream for non-farmer owners, something that doesn't happen if one owns gold or silver, What does this mean for our food system?
>
> It means that farmland is prohibitively expensive for young, beginning farmers. It also means that farmers are getting older. The average age of a farmer in the United States is fifty-eight.
>
> (Holt-Gimenez, 2017)

It is no coincidence that these four settlements occurred under President Barack Obama's administration because the Executive Branch was intentionally made more diverse and representational which allowed for marginalized voices to finally be heard. In spite of the immediate resistance to his policies, President Obama managed to negotiate Congressional approval of all of the settlements for the four farmer groups and even the difficult process of those settlements being actually realized. There were many impediments discovered during this process on both the national and local level but because these systems are so inherently political they are subsequently slow to change.

In the midst of these settlements, which included loan forgiveness was a burgeoning food justice movement that had multiple components along with some cultural and practical changes happening with the U.S.D.A., in the form of better outreach services at both the national and local level. At this same moment we have a global food system that, due to trade agreements – exacerbates the differences between large agribusiness-based farms that have the capacity to compete versus small farms, which make up the bulk of family farms. So the convergence of these challenges and the history of agriculture development in this country served as the motivation for a movement of food justice that put farming access as central to resolving economic disparities.

The U.S.D.A. has to consider how their policies and regulations have served as impediments now that they are working to become a more equitable organization in the global environment. Even how the U.S.D.A. defines farmers and determines who is worthy of access to resources and information can be misguided, especially when stimming from a history of exclusionary practices.

"Despite its pervasiveness, racism is almost never mentioned in international programs for food aid and agricultural development. Although anti-hunger and food security programs frequently cite the shocking statistics, racism is rarely identified as the cause of inordinately high rates of hunger, food insecurity pesticide poisoning, and diet-related disease among people of color. Even the widely hailed "good food movement," with its plethora of project for organic agriculture, permaculture, healthy food, community supported agriculture, farmers' markets, and corner store conversions, tends to address the issue of racism unevenly. Some organizations are committed to dismantling racism in the food system and make this central to their activities. Others are sympathetic but not active on the issue."

(Holt-Gimenez, 2017)

There has been a considerable expansion of local and regional food systems largely because of the efforts toward sustainable agricultural development and a focus on increasing the numbers of young farmers. The need for this focus has become urgent as the average age of United States farmers has increased from one Census of Agriculture to the next (USDA – National Agricultural Statistics Service – Census of Agriculture, n.d.). But this emphasis on increasing the numbers has aligned with the focus on food deserts, access and sovereignty, environmental and conservation movements, and discussions surrounding organics and slow food movements. Some unintended consequences of these efforts are that if there are shortfalls to other local systems, such as public transportation or affordable housing, for example, these efforts sometimes feed into the elitist food system which causes more harm. Therefore, we must consider the sustainability of these young farmers in both our local and global food systems.

The process of becoming a successful farmer has found some answers in the struggles of the marginalized farmers that resorted to forming collectives and cooperatives of mutual support and creating a market for their goods, all prior to the existence of technology and social media. Young farmers today have the skills to and must use technology to access consumers directly but also to exchange information among one another and find creative farming techniques that allow them to maximize their output on relatively small acreage.

What our experience with the COVID-19 pandemic has made clear with the resulting disruptions in the global food systems, is that our reliance upon local producers is even more critical. The innovation of young farmers does not excuse the U.S.D.A. from continuing outreach efforts, what is really necessary today is high-level government support that will help expand the capacity for local farmers to be productive. The U.S.D.A. must also continue to provide oversight into the practices at the local level which first led to the discrimination and maintain its civil rights division.

Current economically driven agricultural practices have some deep inherent flaws that result in degradation of natural systems and human health. As we proceed in our attempts to resolve major ecological rises, individual food producers will have to create ways to survive in the face of new threats to their food production, while at the same time holding firmly to their deepest values. If we are able to expand our view and our response, values of sustainability and stewardship will hopefully become the bottom line.

(Ewing & Gupta, 2019)

References

Ewing, B. B. & Gupta C. 2019. *Nourish: The revitalization of foodways in Hawai'i*. San Francisco, CA: Extracurricular Press.

Holt-Gimenez, E. 2017. *A foodie's guide to capitalism: Understanding the political economy of what we eat*. New York, NY: Monthly Review Press.

"USDA – National Agricultural Statistics Service – Census of Agriculture." Retrieved February 19, 2019, from https://www.nass.usda.gov/AgCensus/.

Bibliography

2007 Census of Agriculture—History (Volume 2, Subject Series Part 7). (2007). United States Department of Agriculture/National Agriculture Statistics Service. https://www.census.gov/history/pdf/2007aghistory.pdf.

2019 County Committee Elections—Farm Services Agency. (2019).

About the U.S. Department of Agriculture | USDA. (n.d.). Retrieved February 1, 2019, from https://www.usda.gov/our-agency/about-usda.

About WFAN – Women, Food and Agriculture Network. (n.d.). *Mission of WFAN.*

Alabama: Selma to Montgomery National Historic Trail. (n.d.). Natioal Park Service.

Alexander, K., American Law Division; Ross W. Gorte, Resources, Science, and Industry Division. (2007). Federal land ownership: Constitutional authority and the history of acquisition, disposal, and retention note. [i]–12.

Alkon, A. H. (2012). *Black, white, and green: Farmers markets, race, and the green economy.* Athens, GA: University of Georgia Press.

Barbour, T. C. (2018). A Community's collective courage: A local food cooperative's impacton food insecurity, community and economic development and climate change. *Butler Journal of Undergraduate Research, 4.*

Bennett, D. (2010). Keepseagle discrimination settlement: $680 million. Fresno, CA: *Western Farm Press.* http://search.proquest.com/docview/847388773/abstract/561F158B9F9E4AD9PQ/19.

Bennett, D. (2011). *Love v Vilsack: Women farmers allege USDA discrimination.* Clarksdale, MS: Southwest Farm Press;. http://search.proquest.com/docview/874476157/abstract/7A8D19FC69824D7CPQ/1

Bigelow, D., Borchers, A., & Hubbs, T. (2016). *U.S. farmland ownership, tenure, and transfer.* USDA Economic Research Service.

Boyd, J. (1995). History—National Black Farmers Association [Http://www.nationalblackfarmersassociation.org/].

"Written testimony of John W. Boyd, Jr. President, National Black Farmers Association."—"Civil Rights in Light of Pigford v. Glickman", (2005) (testimony of Chabot, Steve).

Cobell v. Salazar. (2009). U.S. Department of the Interior.

Cowan, T. & Feder, J. (2006). *Pigford Case: USDA Settlement of a Discrimination Suit by Black Farmers*, 1–6.

Cowan, T. and Feder J. (2010). *The Pigford Cases: USDA Settlement of Discrimination by Black Farmers* [Congressional Research Service].

Cummings, J. D. (2016). *Emancipation proclamation.* Minneapolis, MN: ABDO Publishing.

Dixon, D. & Hapke, H. (2003). Cultivating discourse: The social construction of agricultural legislation. *Annals of the Association of American Geographers.* 93(1), 142–164.

Dunbar-Ortiz, R. (2014). *An Indigenous Peoples' History of the United States.* Boston, MA: Beacon Press.

Evans, D. (n.d.). *Hispanic farmers seek justice for discriminatory USDA lending practices.* Durham, NC: Institute for Southern Studies.

Ewing, B. B. & Gupta C. (2019). *Nourish: The revitalization of foodways in Hawai'i.* San Francisco, CA: Extracurricular Press.

Feder, J. and Cowan T. (2013). *Garcia v. Vilsack: A Policy and Legal Analysis of a USDA Discrimination Case.* Congressional Research Service.

Female Producers. (n.d.). 2017 Census of Agriculture.

Food Sovereignty and the Rights of Indigenous Peoples. (2013). International Indian Treaty Council.

Fullerton, M. (ed). (1992). *What ever happened to Berkeley Co-Op: A collection of essays.* The Center for Cooperatives.

Garvey, T. (2010). *The Indian Trust fund litigation: An overview of Cobell v. Salazar.* Congressional Research Service.

Gottlie, R. & Joshi A. (2010). *Food Justice.* Cambridge, MA: MIT Press.

Hinson, W. R. (2018). Land gains, land losses: The odyssey of African Americans since reconstruction. *American Journal of Economics and Sociology,* 77(3–4), 893–939. https://doi.org/10.1111/ajes.12233

Hispanic Producers. (n.d.).

Holt-Gimenez, E. (2017). *A foodie's guide to capitalism: understanding the political economy of what we eat.* New York, NY: Monthly Review Press.

Ives, S. (1996). *The West: The Geography of hope* [Documentary].

Jablow, V. & Horne B. (1999). Farmers' markets. *Smithsonian Magazine.*

Kendi, Ibram X. (2019). *How to be an antiracist.* New York, NY: One World.

Kenney, T. (2017). 6 successful farms promoting the resurgence of Black Agrarians. *Atlanta Black Star.* https://atlantablackstar.com/2017/05/30/6-successful-black-owned-farms-promoting-resurgence-black-agrarians/4/

Koenig, S. & Dodson, C. (1999). *FSA credit programs target minority farmers* (Agricultural Outlook).

Krom, C. (2010). The real story of racism at the USDA. *The Nation.*

Lincoln's Milwaukee Speech | National Agricultural Library. (n.d.). Retrieved July 8, 2019, from https://www.nal.usda.gov/lincolns-milwaukee-speech

MacDonald, J. M., Hoppe, R. A., & Newton, D. (n.d.). *Three decades of consolidation in U.S. agriculture.* Retrieved February 11, 2019, from http://www.ers.usda.gov/publications/pub-details/?pubid=88056

Mason, D. (2009). *The end of the American century.* Lanham, MA: Rowman & Littlefield Publishers, Inc.

Obstruction of justice: USDA settlement fails Black farmers. (2004). Envronmental Working Group (EWG).

O'Hara, J. (2011). *Market forces: Creating jobs through public investment in local and regional food systems.* Union of Concerned Scientists.

Robinson, J. (2012). *Women, Hispanic farmers say discrimination continues in settlement.* Northwest NEWS Network.

Rodriguez, S. (ed.). (2018). *Food justice: A primer.* Sanctuary Publishers.

S.J. Res.14—A joint resolution to acknowledge a long history of official depredations and ill-onceived policies by the Federal Government regarding Indian tribes and offer an apology to all Native Peoples on behalf of the United States. (2009). 111th Congress (2009-2010). https://www.congress.gov/bill/111th-congress/senate-joint-resolution/14/text.

Stack, C. (1996). *Call to home: African Americans reclaim the rural south.* New York, NY: Basic Books.

Stevens, K. (2007). Gee's Bend. In *Encyclopedia of Alabama.*

Sturgis, S. (2010). *"Still waiting for justice, Black farmers rally in North Carolina.* Durham, NC: Institute for Southern Studies.

Taylor, D. E. (2018). Black farmers in the USA and Michigan: Longevity, Empowerment, and Food Sovereignty. *Journal of African American Studies, 2, 49–76.* https://doi.org/10.1007/s12111-018-9394-8.

Thackeray, L. (2010). Discrimination settlement cold mean payments for Montana Indian farmers. *Billings Gazette.*

The History. (n.d.). Auburn University. Retrieved February 1, 2019, from http://www.auburn.edu

Tribal Food Sovereignty definition. (n.d.). [Wellforculture.com].

U.S. Department of Agriculture: Progress toward implementing GAO's civil rights recommendations. (2012). U.S. Govt. Accountability Office.

USDA – National Agricultural Statistics Service—Census of Agriculture. (n.d.). Retrieved February 19, 2019, from https://www.nass.usda.gov/AgCensus/.

USDA Defines Food Deserts | American Nutrition Association. (n.d.). Retrieved February 11, 2019, from http://americannutritionassociation.org/newsletter/usda-defines-food-deserts.

USDA ERS – Food Access Research Atlas. (n.d.). Retrieved February 11, 2019, from https://www.ers.usda.gov/data/fooddesert/.

USDA History Collection Introduction/Index | Special Collections. (n.d.). Retrieved February 1, 2019, from https://specialcollections.nal.usda.gov/usda-history-collection-introductionindex.

Using SNAP at Farmers' Markets. (n.d.). USDA. https://snaped.fns.usda.gov/nutrition-education/nutrition-education-materials/farmers-markets.

Vilsack, T. (2010). *MY USDA: A progress report for employees on USDA's cultural transformation—Summary of progress November 2010 through February 2011.* USDA Office of the Secretary.

Waddell, B. (2020). "Men like you weren't meant to own land'. *High Country News.*

Washington, Dubois, & Johson. (1901). *Three Negro Classics: Up From Slavery (Booker T. Washington); The Souls of Black Folk (W.E.B. DuBois) and The Autobiography of an Ex-Colored Man.* New York, NY: Avon Books.

White, M. (2018). *Freedom farmers: Agricultural Resistance and the Black freedom movement.* Chapel Hill, NC University of North Carolina Press.

Wilkins, D. & Stark, H. K. (2017). *American Indian politics and the American political system.* Spectrum Series. Lanham, MD: Rowman & Littlefield Publishers.

Index